Modelling for Coastal Hydraulics and Engineering

Modelling for Coastal Hydraulics and Engineering

K.W. Chau

Spon Press
an imprint of Taylor & Francis

LONDON AND NEW YORK

First published 2010
by Spon Press
2 Park Square, Milton Park, Abingdon, Oxon OX14 4RN

Simultaneously published in the USA and Canada
by Spon Press
270 Madison Avenue, New York, NY 10016, USA

*Spon Press is an imprint of the Taylor & Francis Group, an
informa business*

© 2010 K.W. Chau

Typeset in Sabon by
Integra Software Services Pvt. Ltd, Pondicherry, India
Printed and bound in Great Britain by
CPI Antony Rowe, Chippenham, Wiltshire

This publication presents material of a broad scope and applicability.
Despite stringent efforts by all concerned in the publishing process,
some typographical or editorial errors may occur, and readers are
encouraged to bring these to our attention where they represent
errors of substance. The publisher and author disclaim any liability,
in whole or in part, arising from information contained in this
publication. The reader is urged to consult with an appropriate
licensed professional prior to taking any action or making any
interpretation that is within the realm of a licensed professional
practice.

British Library Cataloguing in Publication Data
A catalogue record for this book is available from the British Library

Library of Congress Cataloging-in-Publication Data
Modelling for coastal hydraulics and engineering / [edited by]
K.W. Chau.
p. cm.
Includes bibliographical references and index.
1. Coastal engineering--Mathematics. 2. Coastal engineering--Data
processing. 3. Hydraulic engineering--Mathematics.
4. Coasts--Computer simulation. 5. Coast changes--Mathematical
models. I. Chau, K. W. (Kwok Wing)
TC209.M58 2010
627'.580151--dc22   2009027718

ISBN 10: 0-415-48254-2 (hbk)
ISBN 10: 0-203-88476-0 (ebk)

ISBN 13: 978-0-415-48254-7 (hbk)
ISBN 13: 978-0-203-88476-8 (ebk)

Contents

1   Introduction

In the analysis of the coastal water process, numerical models are often employed to simulate flow and water quality problems. The rapid development of computing technology has furnished a large number of models to be employed in engineering or environmental problems. To date, a variety of coastal models are available, and the modelling techniques have become quite mature. The numerical technique can be based on the finite element method (Kliem *et al.* 2006; Jones and Davies 2007), finite difference method (Buonaiuto and Bokuniewicz 2008; Tang *et al.* 2009), boundary element method (Karamperidou *et al.* 2007; Duan *et al.* 2009), finite volume method (Aoki and Isobe 2007; Qi *et al.* 2009), or Eulerian-Lagrangian method (Cheng *et al.* 1984). The time-stepping algorithm can be implicit (Holly and Preissmann 1977), semi-implicit (Ataie-Ashtiani and Farhadi 2006), explicit (Ghostine *et al.* 2008), or characteristic-based (Ataie-Ashtiani 2007; Perera *et al.* 2008). The shape function can be of the first order, second order, or a higher order. The modelling can be simplified into different spatial dimensions, i.e. a one-dimensional model (Chau and Lee 1991a; Abderrezzak and Paquier 2009), two-dimensional depth-integrated model (Leendertse 1967; Tang *et al.* 2009), two-dimensional lateral-integrated model (Wu *et al.* 2004; Elfeki *et al.* 2007), two-dimensional layered model (Chau *et al.* 1996; Tucciarelli and Termini 2000), three-dimensional model (Blumberg *et al.* 1999; Chau and Jiang 2001, 2002; Carballo *et al.* 2009), and so forth. An analysis of coastal hydraulics and water quality often demands the application of heuristics and empirical experience, and is accomplished through some simplifications and modelling techniques according to the experience of specialists (Yu and Righetto 2001). However, the accuracy of the prediction is to a great extent dependent on open boundary conditions, model parameters, and the numerical scheme (Martin *et al.* 1999)

The adoption of a proper numerical model for a practical coastal problem is a highly specialized task. Ragas *et al.* (1997) compared eleven UK and USA water quality models utilized to find the allowable levels and types of discharge and concluded that model selection was a complicated process of matching model features with the particular situation. These predictive tools inevitably involve certain assumptions and/or limitations, and can be applied only by experienced engineers who possess a comprehensive under-standing of the problem domain. This leads to severe constraints on the use of models, and large gaps in understanding and expectations between the developers and practitioners of a model.

Over the past two decades, there has been a widespread interest in the field of artificial intelligence (AI) (Abbott 1989; Chau 1992a; Abbott 1993; Garrett 1994; Chau and Zhang 1995; Chau and Ng 1996; Ragas *et al.* 1997; Recknagel *et al.* 1997; Maier *et al.* 2001; Chau and Cheng 2002; Chen and Mynett 2003; Chau 2006b; Kalra and Deo 2007; Muttil and Chau 2007; Chen *et al.* 2008; Preis and Ostfeld 2008; Schories *et al.* 2009). The recent advance in AI technologies are making it possible to incorpo-rate machine learning capabilities into numerical modelling systems so as to bridge the gaps between developers and practitioners of a model, and lessen the burdens on human experts. The development of these intelligent management systems is facilitated by employing some shells under the estab-lished development platforms such as MathLab, Visual Basic, C++, and so forth. Owing to the complexity of the numerical simulation of flow and/or water quality, there is an increasing demand to couple AI with these math-ematical models in order to cover more and more characteristics contained in advanced computer technology.

In this book, the development and current progress of integration of dif-ferent AI technologies into coastal modelling are reviewed and discussed. The algorithms and methods investigated include knowledge-based sys-tems (KBS) (Chau 2006; Schories *et al.* 2009), genetic algorithms (Chen *et al.* 2008; Preis and Ostfeld 2008), genetic programming (Kalra and Deo 2007; Muttil and Chau 2007), artificial neural networks (Recknagel *et al.* 1997; Chau and Cheng 2002), and fuzzy inference systems (Maier *et al.* 2001; Chen and Mynett 2003). KBSs have apparent advantages over the other systems in facilitating more transparent transfers of know-ledge in the use of models and in providing intelligent manipulation of calibration parameters. This book may furnish some useful advice to inex-perienced engineers on how to establish a numerical model, although an understanding of the underlying theories is still necessary.

2 Coastal modelling

2.1 Introduction

In this chapter the derivation of the governing equations of hydrodynamics and mass transport in coastal modelling is introduced. The time-varying coastal problem in general is expressed in terms of partial differential equations with the partial derivatives in both temporal and spatial domains. Moreover, the assumptions involved are presented.

2.2 Hydrodynamic modelling

For a fluid which is incompressible and Newtonian, the following equations of motion can be stated (Batchelor 1967; Chau and Jin 1998).

Conservation of mass:

$$u_{i,i} = 0 \tag{2.1}$$

Conservation of momentum:

$$u_{i,t} + (u_i u_j)_{,j} + f_i = \frac{1}{\rho}\sigma_{ij,j} \tag{2.2}$$

Here u_i and f_i are the i-th component of the velocity and body force vectors respectively, in the Cartesian system of coordinated $x_i (i = 1, 2)$; t is the time variable, ρ is the fluid density, $'j = \partial/\partial x_j$ and σ_{ij} is the stress tensor. Expressing the stress tensor in terms of the pressure and the rate of strain tensor, equation (2.2) becomes the Navier–Stokes equations

$$u_{i,t} + (u_i u_j)_{,j} + \frac{1}{\rho}p_{,i} + f_i = \upsilon u_{i,jj} \tag{2.3}$$

with υ being the kinematic viscosity of the fluid and p being the pressure.

For geophysical flows, f_i contains the acceleration due to gravity and the Coriolis force. Geophysical flows, and real engineering flows in general, are almost invariably turbulent and have scales of motion spanning a range of several orders of magnitude. In most engineering problems, it is not necessary to know the exact fine structure of the flow. Only the average values and the overall effects of turbulent fluctuations are studied. The most commonly accepted averaging process is that of Reynolds, who introduced the concept of first replacing an instantaneous value with a mean and fluctuating component and then taking an average over time.

Inserting decompositions for u_i and p into the Navier–Stokes equations, and averaging over a time interval T gives the Reynolds equation

$$u_{i,t} + (u_i u_j)'_j + \frac{1}{\rho} p'_i + f_i = \frac{1}{\rho} \tau_{ij,j} \tag{2.4}$$

where $\tau_{ij} = 2\mu s_{ij} + q_{ij}$, μ is the dynamic viscosity of the fluid, s_{ij} is the rate of strain tensor and q_{ij} is the Reynolds stress tensor.

Geophysical flows of the type to be considered later are nearly horizontal and without appreciable vertical acceleration. The neglect of all dynamic processes such as local acceleration, advective acceleration and all diffusive effects in the vertical direction gives the assumption of hydrostatic pressure distribution, which is common to all shallow water formulations. Thus the momentum equation for the vertical direction $(i = 3)$, considering a right-hand set of axes fixed to the earth and orientated as shown in Figure 2.1, becomes simply

$$\frac{1}{\rho} p'_3 + g = 0 \tag{2.5}$$

or after integration

$$p = \rho g(\eta - x_3) + p_a \tag{2.6}$$

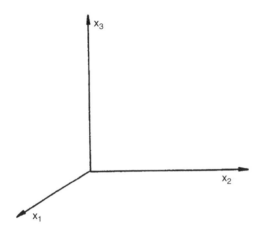

Figure 2.1 Orientation of a right hand set of axes

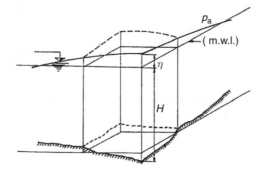

Figure 2.2 Definition sketch for the depth-averaged equations

with reference to the definition sketch of Figure 2.2.

The integration over the vertical water-column now proceeds with the use of Leibnitz's rule (Wylie 1975). The momentum equation is integrated in the i-th direction, term by term. The density ρ has been assumed to be constant.

The depth-averaged velocity components are introduced and are defined by

$$U_i = \frac{1}{(H+\eta)} \int_{-H}^{\eta} u_i \mathrm{d}x_3 \tag{2.7}$$

as well as the equivalence (Kuipers and Vreugdenhill 1973)

$$u_i = U_i + (u_i - U_i) \tag{2.8}$$

to treat the non-linear terms and to retain the same basic equation form for the depth averaged quantities as existed for the time averaged quantities.

The treatment of the momentum equations, after rearrangement, gives

$$(hU_i)_{,t} + \left[hU_iU_j + \delta_{ij}\frac{1}{2}g(h^2 - H^2) \right]_{,j} = hF_i + g(h - H)H_{,i}$$

$$+ \frac{1}{\rho}\left(\tau_{\text{wind}}^i - \tau_{\text{friction}}^i\right) + \frac{1}{\rho}\int_{-H}^{\eta}\left[(\tau_{ij} - \rho(u_i - U_i)(u_j - U_j))\,\mathrm{d}x_3\right]_{,j} \tag{2.9}$$

where $h = H + \eta$. Here, for simplicity, p_a is assumed to be constant; however, the inclusion of an atmospheric pressure depending on the location $p_a = p_a(x)$ can be readily accommodated.

The only equation left which has to be integrated through the depth is the continuity equation (2.1). Carrying out vertical integration term by term, and invoking again Leibnitz's Rule, we get

$$\int_{-H}^{\eta} u_{i,i} dx_3 = 0 \tag{2.10}$$

Applying the kinematic boundary conditions and inserting depth averaged velocity components, we finally get

$$h_{,i} + (hU_i)_{,t} = 0 \tag{2.11}$$

2.3   Water quality modelling

The basic governing equation of general mass transfer or convection–diffusion is

$$(\rho c)_{,t} + (\rho u_i c)_{,i} = [E_i (\rho c)_{,i}]_{,i} + r_i \tag{2.12}$$

where E_i are horizontal diffusion coefficients and r_i is the net internal input rate (Fischer and List 1979).

Integrating each term over the vertical from the bottom to the free surface and utilizing Leibnitz's Rule, we obtain the following equation:

$$\frac{\partial}{\partial t}(hc) + \frac{\partial}{\partial x_i}(hU_i c) = \frac{\partial}{\partial x_i}\left(hD_{ij}\frac{\partial c}{\partial x_i}\right) - khc + S \tag{2.13}$$

where c is the depth-averaged concentration, D_{ij} is the dispersion coefficient, k is the decay coefficient and S is the source term.

2.4   Governing equations

Ignoring the effect of the longitudinal turbulent shear stresses, wind stress and the Coriolis force, the two-dimensional governing equations for tidal hydrodynamics and mass transport are:

Conservation of mass:

$$\frac{\partial h}{\partial t} + \frac{\partial (hu)}{\partial x} + \frac{\partial (hv)}{\partial y} = 0 \tag{2.14}$$

Conservation of x-momentum:

$$\begin{aligned}
\frac{\partial (hu)}{\partial t} + \frac{\partial}{\partial x}\left[hu^2 + \frac{1}{2}g(h^2 - H^2)\right] + \frac{\partial}{\partial y}(huv) = g(h - H)\frac{\partial H}{\partial x} \\
- \frac{gu}{C_z^2}(u^2 + v^2)^{1/2} + \frac{\partial}{\partial x}\left[\frac{2\mu h}{\rho}\left(\frac{\partial u}{\partial x}\right)\right] + \frac{\partial}{\partial y}\left[\frac{\mu h}{\rho}\left(\frac{\partial v}{\partial x} + \frac{\partial u}{\partial y}\right)\right]
\end{aligned} \tag{2.15}$$

Conservation of *y*-momentum:

$$\frac{\partial(hv)}{\partial t} + \frac{\partial}{\partial x}(huv) + \frac{\partial}{\partial y}\left[hv^2 + \frac{1}{2}g(h^2 - H^2)\right] = g(h - H)\frac{\partial H}{\partial y}$$

$$- \frac{gv}{C_z^2}(u^2 + v^2)^{1/2} + \frac{\partial}{\partial x}\left[\frac{\mu h}{\rho}\left(\frac{\partial v}{\partial x} + \frac{\partial u}{\partial y}\right)\right] + \frac{\partial}{\partial y}\left[\frac{2\mu h}{\rho}\left(\frac{\partial v}{\partial y}\right)\right] \quad (2.16)$$

Scalar transport:

$$\frac{\partial}{\partial t}(hc) + \frac{\partial}{\partial x}(huc) + \frac{\partial}{\partial y}(hvc) = \frac{\partial}{\partial x} \geq \left(hD\frac{\partial c}{\partial x}\right) + \frac{\partial}{\partial y}\left(hD\frac{\partial c}{\partial y}\right) \quad (2.17)$$

in which u and v are the depth-averaged components of the velocity in the coordinate directions x and y, h is the total height of fluid, H is the depth measured with respect to the mean water level, g and ρ are the gravity acceleration and density respectively, C_z and μ are the Chezy coefficient and eddy viscosity coefficient, c is a scalar representing pollutant concentration and D is the coefficient representing combined diffusion and dispersion in the x and y directions.

2.5 Conclusions

In this chapter the derivation of the governing equations of hydrodynamics and mass transport in coastal modelling has been introduced. Moreover, the assumptions involved have been presented. In the next chapter, we will discuss the conventional numerical tools used to address coastal engineering problems. The notion of "generations" of modelling to describe the trend of development will be introduced. The feasibility of the incorporation of artificial intelligence techniques into contemporary modelling will be discussed. Several common features for different numerical discretization methods to a simple partial differential equation will be highlighted. Some basic differences between these numerical methods will also be mentioned.

3 Conventional modelling techniques for coastal engineering

3.1 Introduction

In this chapter, we will discuss the conventional numerical tools used to address coastal engineering problems. The notion of "generations" of modelling to describe the trend of development will be introduced. The feasibility of the incorporation of artificial intelligence techniques into contemporary modelling will be discussed. Several common features for different numerical discretization methods to a simple partial differential equation will be highlighted. Some basic differences between these numerical methods will also be mentioned.

3.2 Mechanistic modelling

Numerical modelling can be defined as a process that transforms knowledge on physical phenomena into digital formats, simulates for the actual behaviours, and translates the numerical results back to a comprehensible knowledge format (Abbott 1989). In mechanistic models, the equation for the transport of pollutants can be expressed as:

$$\frac{\partial SD}{\partial t} + \frac{\partial SUD}{\partial x} + \frac{\partial SVD}{\partial y} + \frac{\partial S\omega}{\partial \sigma} = \frac{\partial}{\partial x}\left(A_s H \frac{\partial S}{\partial x}\right) + \frac{\partial}{\partial y}\left(A_s H \frac{\partial S}{\partial y}\right)$$
$$+ \frac{\partial}{\partial \sigma}\left[\frac{K_H}{D}\frac{\partial S}{\partial \sigma}\right] - K_s DS + S_s \tag{3.1}$$

where (U, V, ω) are mean fluid velocities in the (x, y, σ) directions; S is the density of the pollutant; $D = \eta + H$, where η is the elevation of the sea surface above the undisturbed level, and H is the undisturbed mean depth of the water; and K_H is the vertical turbulent flux coefficient, which can be derived from the second moment $q^2 \sim q^2 l$ turbulence energy model (Chau 2004a). K_s is the decay rate of pollutant; S_s is the source of the pollutant; and A_s is the horizontal turbulent coefficient. Pollutant transport equations can then be written in discretized forms, depending on which algorithm is used.

3.2.1 Model manipulation

Model manipulation is always entailed, particularly during the initial establishment of the model, since quite different results might arise from a slight
change in the parameters. The procedure is a mixture of feedback and
modification. Knowledge of model manipulation comprises real physical
observations, numerical representation of water movement or water quality,
the discretization of governing equations for physical and chemical processes, schemes to solve the discretized equations effectively and accurately,
and an analysis of output. Experienced modellers can reason from the failure of a model according to a comparison of the modelling results with real
data, as well as a heuristic judgement of key environmental behaviour. The
knowledge mentioned in the above may be employed unconsciously. However, many model users do not have the requisite knowledge to gather their
input data, establish algorithmic models, and assess their results. The result
may be inferior designs, resulting in the underutilization, or sometimes even
the total failure, of these models.

The ultimate goal of model manipulation in coastal engineering is to
obtain satisfactory modelling. As such, a balance should be found between
modelling accuracy and speed. It is observed that modellers usually maintain
certain fundamental parameters unaltered during the manipulation process. For example, when researchers were accustomed to two-dimensional
coastal modelling, they altered only the bottom friction coefficient (Chau
and Jin 1995). In water quality modelling, Baird and Whitelaw (1992)
proposed that the algal behaviour was associated closely not only to its
respiration rate but also the water temperature. Model users would then
consider variations in the intensity of sunlight within the water column
when modelling the phenomenon of eutrophication (Chau and Jin 1998).
These examples indicated that human intelligence employed existing knowledge to decrease the number of selections so as to enhance the effectiveness
of model manipulation. Each time, they tend to vary a minimum number
of parameters. This is understandable because if they make adjustment to
many parameters simultaneously, they may easily get lost as to the direction
of the manipulation. To this end, artificial intelligence techniques are capable of mimicking a more comprehensive process as well as of complementing
the deficiency of human reasoning.

3.2.2 Generations of modelling

The notion of "generations" of modelling to delineate the trend of development was introduced by Abbott (1989) and Cunge (1989). The so-called
third generation modelling is a system to solve specific domain problems.
It is only intelligible to the modeller and special users well-trained over
a long period. It has included very few features to facilitate understanding by other users or to handle other user-friendly interface problems.

Typical examples are some sophisticated convection–dispersion models of the Eulerian-Lagrangian type (Cheng *et al.* 1984), two-dimensional or three-dimensional finite difference numerical models on tidal flow (Chau *et al.* 1996; Blumberg *et al.* 1999) and on a specific water quality phenomenon such as eutrophication (Chau and Jin 1998), finite element modelling of floodplain flow (Tucciarelli and Termini 2000), the depth-averaged turbulence k-e model (Yu and Righetto 2001), etc.

Some previous efforts have been devoted to coping with a much wider range of end-users. The fourth generation of modelling has become much more meaningful to a much wider range of end-users. It furnishes a menu of parameter specifications, automatic grid formation, preprocessing and post-processing features, and features for the management of real gleaned data for simulation, etc. These tools act as intelligent front-ends to backup the manipulation of the numerical models for specific hydraulic (Knight and Petridis 1992) or water quality (Recknagel *et al.* 1994) problems. Nevertheless, they do not address the core problem of the elicitation and transfer of knowledge. The modern age is characterized by a boom in knowledge, and the fourth generation of modelling triggers the technological research to transform the knowledge of hydrodynamic and water quality computation into the products.

3.2.3   *Incorporation of artificial intelligence (AI) into modelling*

During the past two decades, the general availability of sophisticated personal computers with ever-expanding capabilities has generated increasing complexity in terms of computational ability in the storage, retrieval, and manipulation of information flows. With the recent advances in AI technology, there has been an increasing demand for a more integrated approach, more than just the need for better models. This claim is justified from the fact of relatively low utilization of models in the industry in comparison to the number of reported models. It is anticipated that this increased capability will both add value to the contemporary decision-making tool to users and streamline the coastal planning and control process.

3.3   Temporal and spatial discretizations

Many temporal and spatial discretization methods have been used in the coastal engineering field: the finite difference method (Buonaiuto and Bokuniewicz 2008; Tang *et al.* 2009), finite element method (Kliem *et al.* 2006; Jones and Davies 2007), boundary element method (Karamperidou *et al.* 2007; Duan *et al.* 2009), finite volume method (Aoki and Isobe 2007; Qi *et al.* 2009), method of characteristics (Ataie-Ashtiani 2007; Perera *et al.* 2008), fractional step method (Ataie-Ashtiani and Farhadi, 2006; Abd-el-Malek and Helal 2009), etc. Walters *et al.* (2009) compared a class of unstructured grid models furnishing a flexible spatial discretization where the continuity equation reduced to a finite volume approximation whilst momentum equations were approximated with finite difference, finite element, or finite volume methods, respectively. Results indicated that the performance of each method relied upon the classes of problems to be encountered, namely, hydraulics, coastal, global ocean flows, etc. Amongst these differences, restrictions on grid irregularity and stability of the Coriolis term were more important.

As mentioned above, two widely used discretization techniques are the finite difference method and the finite element method. A basic distinction between them is that the former generates the numerical equations at a given point based on the values at neighbouring points, whilst the latter generates equations for each element independently of all the other elements. It is only when the finite element equations are coupled together and assembled into the global matrices that the interaction between elements is taken into consideration. Another distinction is in the application of the boundary conditions. For the finite difference method, fixed-value boundary conditions can be directly inserted into the solution whilst the discretized equation has to be modified to account for derivative boundary conditions. For the finite element method, the treatment is totally different. Derivative boundary conditions are already considered during the formation of element equations whilst fixed-value boundary conditions have to be applied to the global matrices.

Several common features exist for different numerical discretization methods to a simple partial differential equation. First of all, the governing equations have to be transformed into discretized equations for the values of the variable at a finite number of points in the domain under consideration, using an appropriate temporal and spatial discretization scheme. The form of equations generated can be explicit or implicit. If an implicit form is generated, a set of simultaneous equations have to be solved. Prior to the computation, appropriate boundary conditions have to be set corresponding to the actual situation. A set of initial conditions is involved to start the computation for the time-dependent problem.

When these discrete equations are solved to obtain a set of values for the variables at points in the domain, our basic requirement of the solution is accuracy. The concepts for this in numerical modelling are convergence and stability. Convergence is the ability of a set of numerical equations to represent the analytical or exact solution to a problem, if it exists. A stable solution is accomplished if the errors in the discrete solution do not grow along with the numerical process so that the equation will move towards a converged solution. If a diverging solution is obtained, it is probable that some mistakes have been made. Some possible causes are:

- a poor spatial mesh not matching with the orthogonal nature of the scheme;
- inappropriate representation of the boundary conditions;
- improper/unrealistic representation of the initial conditions;

- insufficient upwinding for the convection terms; and
- inappropriate representation of the turbulence model.

Cheng *et al.* (1984) analysed a Eulerian-Lagrangian method (ELM) of solution for the convection–dispersion equation and for treatment of anisotropic dispersion in natural coordinates, in which the Lagrangian concept was employed in a Eulerian computational grid system. Results indicated that the use of second-order Lagrangian polynomials in the interpolation would not generate artificial numerical dispersion.

Chau and Jin (1995) presented a two-layered, two-dimensional mathematical model using a finite difference method which could be employed to mimic flows with density stratification in a natural water body with complicated topography. In the model the turbulent exchange across the interface was treated empirically and a time-splitting finite difference method with two fractional steps was used to solve the governing equations. The model was calibrated and verified by comparing the numerical results with data measured in Tolo Harbour, Hong Kong. The numerical results simulated the field measurements very closely. The computation indicated that the model simulated the two-layer, two-dimensional tidal flow with density stratification in Tolo Harbour very well. The computed velocity hodographs illustrated that the tidal circulations at various positions in each layer had different patterns and that the characteristics of the patterns were not related to the type of the tide except for their scales. The computed Lagrangian pathlines demonstrated that the tidal excursion was related to the tidal type, especially in the inner harbour and side coves.

Chau *et al.* (1996) implemented an unsteady finite difference mathematical model of depth-averaged two-dimensional (2-D) flow for Tolo Harbour in Hong Kong employing the numerically generated boundary-fitted orthogonal curvilinear grid system and the grid "block" technique. The model overcame the Courant–Friedrichs–Lewy stability criterion constraint and provided more freedom in the specification of a flexible initial condition. The error bars and the root mean square errors between the simulated and the field data demonstrated that this model reproduced the depth-averaged 2-D flow in Tolo Harbour reasonably, and the computational results concurred with the available field data. It could be noted from the numerical model that the flow exchange or tidal flushing in the inner part of the harbour and in the small side coves was quite limited.

Flow and transport in a natural water body often interact with density stratification and in such cases the stratification might be characterized as a two-layered system. Chau and Jin (1998) proposed a rigorous, two-layered, two-dimensional finite difference numerical model for eutrophication dynamics in coastal waters using a numerically generated, boundary-fitted, orthogonal curvilinear grid system. The model mimicked the transport and transformation of nine water quality constituents pertinent to eutrophication. The architecture of the model followed a generally accepted framework, with the exception of the interaction between the two layers via vertical advection and turbulent diffusion. Some kinetic coefficients were calibrated with field data specifically for the scenario in Tolo Harbour, Hong Kong. The pollution sources were unsteady, and hourly solar radiation was prescribed. From the *in situ* sampling analysis, sediment oxygen demand and nutrient releases from sediment were included in the model. The hydrodynamic variables were predicted simultaneously with another hydrodynamic model detailed in Chau *et al.* (1996). The simulated results indicated that the model successfully simulated the stratification tendency in all the water quality constituents, demonstrating an obvious bottom water anoxic condition during the summer. The result was consistent with the density stratification and the unsteady layer-averaged 2-D eutrophication processes which were observed in Tolo Harbour.

Blumberg *et al.* (1999) applied the Estuarine, Coastal and Ocean Model with a single curvilinear and orthogonal grid system to conduct three-dimensional simulations of estuarine circulation in the New York Harbor complex, Long Island Sound, and the New York Bight from the semidiurnal to the annual scales. It took into consideration several model-forcing functions, comprising meteorological data, water level elevation, temperature and salinity fields along the open boundary, freshwater inflows, wastewater treatment plants, and point sources from combined sewer overflows. A result comparison was performed with field observation data of water levels, currents, temperature and salinity, from an extensive hydrodynamic monitoring programme.

Tucciarelli and Termini (2000) applied a split methodology for a two-dimensional solution of the diffusive shallow water equations to simulate flood flow on a river floodplain, by splitting the unknowns of the momentum and continuity equations into one kinematic and one parabolic component. The former was determined using the slope of the water level surface computed in the previous time-step and a zero-order approximation of the water head whilst the latter was computed by applying a standard finite element Galerkin procedure.

Yu and Righetto (2001) presented the developments and applications of a turbulence depth-averaged two-equation closure model in order to determine the distribution of turbulence eddy viscosity and diffusivity as well as other physical variables. Results on the computed distributions and variations of velocity, temperature and concentration fields for two unsteady case studies in the Rhine River and in the Yangtse River were compared with those computed by different turbulence two-equation closure models as well as with experimental results and field data.

Lachaume *et al.* (2003) utilized a two-dimensional numerical wave tank model to investigate the shoaling, breaking, postbreaking of waves, and the transformation of solitary waves over plane slopes. It coupled a boundary element model and a volume of fluid model in addressing potential flow equations and Navier–Stokes equations, respectively. The salient properties of waves breaking over various slopes, such as shape, internal velocities and type of breaking, were computed. A comparison of the results was made against existing laboratory experiments.

Chau (2004a) delineated a real-time three-dimensional finite difference numerical model for eutrophication dynamics in coastal waters of Tolo Harbour, Hong Kong. The difficulties encountered and subtleties of the modelling as well as some principal kinetic parameters were highlighted. Long-term time-series modelling results due to time-varying pollution sources in Tolo Harbour were also detailed.

Umeyama and Shintani (2004) computed the wave profiles during the run-up and breaking of internal waves over a sloping boundary by using the method of characteristics. Comparison was made against laboratory experiments conducted in a 2-D wave tank, in which a fluid consisting of fresh water and salt water was mixed to imitate the density-stratified ocean. Moreover, a set of luminance data analysed by an image processing technique was employed to reveal the profile of internal waves and the mixture of the upper-layer and lower-layer water and to compare with the predicted density variation as a result of the mixing process.

Kliem *et al.* (2006) developed and implemented a two-dimensional shallow water model with an unstructured mesh for simulation of tide and storm surges in the North Sea/Baltic Sea. Three simulations were made: a ten-day simulation of the M_2 tide only, a one-year full tidal simulation, and a one-year predictive simulation including both tides and atmospheric forcing. The results of sea level predictions were compared with those of a benchmarking finite difference model.

Ataie-Ashtiani and Farhadi (2006) utilized a meshless moving-particle semi-implicit (MPS) method with a fractional step method of discretization to split each time step into two steps to simulate incompressible inviscid flows with free surfaces in a typical dam-break problem. The motion of each particle was computed through interactions with neighbouring particles covered with a variety of kernel functions. The kernel function having the ability to improve the stability of the MPS method was determined by numerical experiments. Results illustrated the advantage of this method in furnishing an accurate prediction of the free water surface for coastal engineering problems involving an irregular free surface.

Jones and Davies (2007) employed a finite element model to determine the spatial distribution of the tides with different tidal harmonic constituents, namely, M_2, S_2, N_2, K_1, O_1, M_4 and M_6, in the west coast of Britain. Comparisons of the results were made against both field observations and a standard finite difference model. Results indicated that the refinement of detailed topography and associated mesh sizes in near-coastal regions by the finite element model contributed to an improvement in accuracy over the benchmarking model.

Aoki and Isobe (2007) presented a one-way nested model with a structured finite difference Princeton Ocean Model (POM) for the

Tsushima-Korea Straits, and an unstructured finite volume model for Fukuoka Bay divided into triangular-cell grids in hindcasting the wind-induced sea-level variations at Hakata tidal station in winter. A result comparison was made against the benchmarking finite difference POM for both the Tsushima-Korea Straits and Fukuoka Bay. It was shown that the nested model constructed with structured and unstructured models outperformed the benchmarking model.

Ataie-Ashtiani (2007) developed a quasi-three-dimensional numerical model, by using the method of characteristics to solve the advection–dispersion equation, for quantification of groundwater flow and pertinent contaminant transport discharged from layered coastal aquifers into the coastal zone. A variety of feasible scenarios were modelled. It was found that the amount of discharged contaminant was highly dependent on the value of hydraulic conductivity. Moreover, results indicated that over-simplification of the seaward boundary condition, in numerical simulations, might cause an incorrect estimate of temporal and spatial variations of the discharged contaminant into coastal water.

Karamperidou *et al.* (2007) undertook a preliminary study of seawater intrusion problem in the coastal aquifer of Eleftherae-N. Peramos, Greece, by employing a numerical water flow model coupling boundary elements and moving points. The effects of some simplifying assumptions made in the numerical simulation tool on the accuracy of the results were presented. Evaluation was also made of some possible scenarios with lower water demand. Some methods to alleviate the local seawater intrusion problem were highlighted.

Gilbert *et al.* (2007) reported on the use of a numerical wave tank, based on a higher-order boundary element method, with an explicit second-order time stepping, in driving simulations of flow and sediment transport around partially buried obstacles. The suspended sediment transport was modelled in the near-field in a Navier–Stokes model employing an immersed-boundary method and an attached sediment transport simulation module whilst turbulence was represented by large eddy simulation. Applications were presented for both single frequency waves and modulated frequency wave groups.

Buonaiuto and Bokuniewicz (2008) employed a finite difference model on the basis of the wave action balance equation to compute wave-driven currents and in turn to investigate the intermittent movement of sediment throughout the Shinnecock Inlet ebb shoal complex. In order to determine the distribution of hydrodynamic forces and investigate the dominant pattern of morphology change, various combinations of incident waves and tide were simulated in the inlet modelling system for three configurations of Shinnecock Inlet at three different dates.

Daoud *et al.* (2008) developed a two-dimensional implicit finite volume scheme for modelling the flow pattern in a closed artificial lagoon and along the coastline near Damietta Port in Egypt, incorporating the effects of the

Coriolis force, surface wind stress, and waves. The scheme employed a non-uniform rectilinear forward-staggered grid with Cartesian coordinates, the Euler implicit technique for time integration, the deferred correction method for treatment of the convective flux, and a second-order central difference approximation for discretization of the viscous terms. Comprehensive validation of the performance of the model was performed under different sources of external forces and different combinations of boundary conditions.

Perera *et al.* (2008) presented a three-dimensional numerical model of the finite-difference method and method of characteristics to solve the partial difference equations of groundwater flow and solute transport and the advection term of mass transport equation, respectively, in order to understand the density-dependent solute transport process taking place in the seawater intrusion problem in the Motooka area of Fukuoka Prefecture, Japan. The transition zone approach, which coupled the groundwater flow and mass transport equation to solve the density dependent flow, was adopted.

Abualtayef *et al.* (2008) delineated the development and application of a three-dimensional multilayer hydrostatic model, coupling the fractional step method, the finite difference method in the horizontal plane and the finite element method in the vertical plane, in order to compute wetting and drying in tidal flats due to tidal motion in Saigo fishery port and the Ariake Sea. Results indicated good agreement with available field observations.

Qi *et al.* (2009) developed and implemented an unstructured-grid finite volume model in order to study the wind-induced surface waves in the Gulf of Maine and New England Shelf, United States, which is a coastal ocean region with complex irregular geometry. The flux-corrected transport algorithm, the implicit Crank–Nicolson method, and options of explicit or implicit second-order upwind finite volume schemes were adopted in frequency space, directional space and geographic space, respectively. It was shown that the performance of the second-order finite volume method was similar to a benchmarking third-order finite difference method.

Tang *et al.* (2009) presented a finite difference discretized numerical model for synchronously coupling wave, current, sediment transport and seabed morphology for the simulation of multi-physics coastal ocean processes. Validation of the results by this model was made against analytical solutions. Moreover, the performance of the numerical model was tested against some typical case studies, namely, simulation of dam-break flow over a mobile-bed and evolution of a wave-driven sand dune. In such cases, the interactions among waves, currents, and seabed morphology were demonstrated.

Duan *et al.* (2009) modelled the initial stage of plunging wave impact obliquely on coastal structures through an oblique collision of an asymmetrical water wedge and an asymmetrical solid wedge, by employing a boundary element method through the Cauchy theorem in the complex plan. Through the numerical experiment, some insights were gained into wave elevation, pressure distribution, forces and moments, effects of different impact angles and effects of oblique impact, etc.

Abd-el-Malek and Helal (2009) applied a fractional steps technique for time-stepping in the numerical solution of the shallow water equations in order to study the water velocity, concentration, temperature distribution, and most important of all, the pollution problem in Lake Mariut, Egypt. The Riemann invariants of the equations were interpolated at each time step along the characteristics of the equations using a cubic spline interpolation. Equations were evolved without the requirement of the iterative steps involved in the multidimensional interpolation problem. The efficient and simple method rendered it appropriate for problems necessitating small time steps and grid sizes and for parallel computing.

3.4   Conclusions

In this chapter, we have briefly discussed the conventional numerical tools used to address coastal engineering problems. The notion of "generations" of modelling to describe the trend of development was introduced. The feasibility of the incorporation of artificial intelligence techniques into contemporary modelling was presented. Several common features for different numerical discretization methods for a simple partial differential equation were highlighted. Some basic differences between these numerical methods were also mentioned. In the following two chapters, the two most widely used computational fluid dynamics tools, namely, finite difference methods and finite element methods, will be delineated with more details. A real prototype application case study will also be introduced.

4 Finite difference methods

4.1 Introduction

This chapter describes finite difference methods often employed in coastal hydraulics in general, and highlights a three-dimensional hydrodynamic and pollutant transport numerical model with an orthogonal curvilinear coordinate in the horizontal direction and a sigma coordinate in the vertical direction. This model is based on the Princeton Ocean Model (POM). In this model a second moment turbulence closure sub-model is embedded, and the stratification caused by salinity and temperature is considered. Furthermore, in order to adapt to estuary locations where the flow pattern is complex, the horizontal time differencing is implicit with the use of a time-splitting method instead of the explicit method in POM. An efficient as well as simple open boundary condition is employed for pollutant transport in this mathematical model. This model is applied to the Pearl River estuary, which is the largest river system in South China with Hong Kong at the eastern side of its entrance. The distribution and transport of chemical oxygen demand (COD) in the Pearl River Estuary (PRE) is modelled. The computation is verified and calibrated with field measurement data. The computed results mimic the field data well, which show that the trans-boundary or inter-boundary effects of pollutants, between the Guangdong Province and the Hong Kong Special Administrative Region due to the wastewater discharged from the Pearl River Delta Region (PRDR), are quite strong.

4.2 Basic philosophy

The finite difference method is most widely used in engineering practice. The foundation of the finite difference method is the following: functions of continuous arguments which describe the state of flow are replaced by functions defined on a finite number of grid points within the computational domain. The original partial derivatives in partial differential equations in temporal and spatial terms represent the time-varying and spatially varying nature of the system. The derivatives are then replaced by divided differences. This method is based upon the use of Taylor series to build a set of equations that describe the derivatives of a variable as the differences between the values of the variable at various points in space or time (Smith 1985). The different ways in which derivatives and integrals are expressed by discrete functions are called finite difference schemes. These difference equations link the values of variables at a set of points to the derivatives. This grid of points is employed to represent the spatial domain throughout the execution of the model. By the application of a finite difference method, the differential equations are reduced to a system of algebraic equations for which two possibilities for solution may be distinguished. These are called explicit and implicit schemes.

In the explicit scheme, the unknown values at a grid point at an instant $(n+1)\Delta t$ are expressed entirely as functions of known data at a number of adjacent grid points at instant $n\Delta t$ (Dronkers 1969, Harleman and Lee 1969); Given the initial conditions, the values at $t = 0$, and the boundary conditions, we can proceed step by step to obtain the grid function for all $t = n\Delta t$. In explicit schemes the system of algebraic equations can be called uncoupled.

In the implicit scheme, the system of algebraic equations is coupled since the spatial derivatives are evidently expressed as a weighted average of the variables at the two time levels (Cunge *et al.* 1980). Because of the coupling between the equations, a simple formula for the solution of individual points cannot be obtained, and a whole set of algebraic equations must be solved simultaneously.

In the following few sections, some aspects of finite difference methods as applied to progressively more complete differential equation descriptions in coastal hydraulics are delineated. For all cases, the basic Navier–Stokes equation system governs completely general flow fields in three-dimensional configurations. In each case, the system is simplified by enforcing different assumptions with different degrees of restrictions on certain physical and/or geometric aspects.

4.3   One-dimensional models

In fact, one-dimensional models are not often used. The advances of computing technology in these few decades mean that the computing effort is no longer a controlling factor. In general, higher accuracy attached to higher dimensional models will be the more deciding factor in the choice of numerical model.

Chau and Lee (1991) implemented an accurate as well as efficient solution to the non-linear de Saint-Venant equation describing unsteady open channel flow. The time history of stages and discharges within individual segments in any channel could be simulated in a connected, essentially one-dimensional network, subject to initial and boundary conditions. The mathematical model developed was based on the four-point operators

Preissmann implicit finite difference scheme. Real hydraulic features, including branched channels and tidal flats, were simulated. The model was subjected to carefully chosen analytical test problems which embraced many essential realistic features of coastal hydrodynamic applications. The model was then applied to study the tidal dynamics and potential flood hazards in the Shing Mun River network, Hong Kong.

Laguzzi *et al.* (2001) applied a robust 1-D finite difference scheme (Delft's scheme) and the Conjugated Gradients method to solve the river channel network. A characteristic of the Delft scheme was its ability to tackle steep fronts, subcritical and supercritical flow. It was demonstrated that dam-break/dike breaks could be readily simulated using the controlled structures from the 1-D system and that the study of flood events on natural river basins, polders, channel-networked regions, urban areas and coastal areas could be facilitated. The results were also compared with those generated by a 2-D rectangular grid hydrodynamic model.

Vennell (2007) presented a 1-D finite difference model to compute the amplitudes of the transient waves generated by a small fast-moving storm crossing a topographic step on to a continental shelf and across a ridge. By applying this model, large transients were generated by storms whose translation speed was faster than the shallow-water wave speed. The paper also discussed the influence of a finite-width shelf on the enhancement of coastal storm surge.

Alho and Aaltonen (2008) illustrated the capability of one-dimensional jökulhlaup simulation (HEC-RAS modelling software) and compared simulation results with those from a two-dimensional model (TELEMAC-2D modelling software) to determine the potential and limitations of one-dimensional modelling for the simulation of extreme glacial outburst floods. They showed that one-dimensional modelling of jökulhlaup propagation provided results broadly comparable to data derived from more complex simulations.

Abderrezzak and Paquier (2009) presented and tested a one-dimensional numerical model for simulating unsteady flow and sediment transport in open channels employing an explicit finite difference scheme. The bed morphodynamics was represented by the Exner equation and an additional equation describing the nonequilibrium sediment transport. The pertinence of the model was examined for two hypothetical cases. Application of the model was made to simulate the morphological changes taking place in the Ha! Ha! River (Quebec) after the failure of the Ha! Ha! Dyke in July 1996.

4.4   Two-dimensional models

Whilst 2-D depth-averaged models are often employed in most cases in coastal engineering, 2-D laterally averaged models will be used when the relationship between the pertinent parameters and the water depth is of great concern.

4.4.1 2-D depth-integrated models

Lin and Chandler-Wilde (1996) developed and applied a 2-D depth-integrated, conformal boundary-fitted, curvilinear model for predicting the depth-mean velocity field and the spatial concentration distribution in estuarine and coastal waters, using the ADI finite difference scheme with a staggered grid. The conformally generated mesh could provide greater detail where it was needed close to the coast, with larger mesh sizes further offshore. It accomplished minimization of the computing effort and maximization of accuracy simultaneously.

Song *et al.* (1999) established a 2-D horizontal plane numerical model for coastal regions of shallow water. Application of the model was made to simulate the circulating flow in the area induced by wind and the tidal flow field of the radial sandbanks in the South Yellow Sea. A key feature of this model was the simultaneous computation of velocity profiles when the equations of the value of difference between the horizontal current velocity and its depth-averaged velocity in the vertical direction were solved. Its performance was compared with that of a quasi-3-D numerical model

Chau and Jin (2002) delineated a robust unsteady two-layered, 2-D finite difference numerical model for eutrophication in coastal waters. The modelling was based upon the numerically generated boundary-fitted orthogonal curvilinear grid system and integrated with a hydrodynamic model. It simulated the transport and transformation of nine water quality constituents associated with eutrophication in the waters, i.e. three organic parameters (carbon, nitrogen and phosphorus), four inorganic parameters (dissolved oxygen, ammonia, nitrite + nitrate and orthophosphate), and two biological constituents (phytoplankton and zooplankton). Key kinetic coefficients were calibrated with the field data. The hydrodynamic, pollution source and solar radiation data in the model were real-time simulated.

Bingham and Agnon (2005) derived a Boussinesq method that was fully dispersive in order to solve for highly non-linear steady waves. Amongst several implementation methods, one of them was the finite difference fast-Fourier transform implementation of the method, which was delineated and applied to more general problems including Bragg resonant reflection from a rippled bottom, waves passing over a submerged bar, and non-linear shoaling of a spectrum of waves from deep to shallow water.

Vennell (2007) developed a 2-D finite difference numerical model in order to illustrate the topographic transients generated by sub- and supercritical storms moving across a ridge. In such case, long surface gravity waves were found to be radiated when storms were crossing topography with topographic transients possibly feeding the required energy. This might be the first one to discuss the generation mechanism for this type of long waves as a result of changes in the depth-dependent amplitude of the atmospherically forced pressure wave beneath a storm. The results were compared to those computed by a 1-D finite difference model.

Tang *et al.* (2009) presented a framework for synchronously coupling nearshore waves, currents, sediment transport, and seabed morphology for the accurate simulation of multi-physics coastal ocean processes. The governing equations were discretized using 2-D finite difference methods. The key characteristics of this framework were the simulation of dam-break flow over a mobile-bed and evolution of a wave-driven sand dune and the illustration of the interactions among waves, currents, and seabed morphology.

4.4.2   2-D lateral-integrated models

Wu *et al.* (2004) developed a semi-implicit shallow water flow numerical model based on the unsteady Reynolds-averaged Navier–Stokes equations with the hydrodynamic pressure assumption. Moreover, the equations were transformed into the sigma-coordinate system and the eddy viscosity was calculated with the standard k-epsilon turbulence model. The model was applied to the 2-D vertical plane flow of a current over two steep-sided trenches for predicting the flow in a channel with a steep-sided submerged breakwater at the bottom.

Liu *et al.* (2005) studied the significance of wave field near structures in coastal and offshore engineering by simulation of the wave profile and flow field for waves propagating over submerged bars. A PLIC-VOF (Piecewise Linear Interface Construction) model was employed to trace the free surface of wave and finite difference method to solve vertical 2-D Navier–Stokes equations. In order to demonstrate that the PLIC-VOF model was effective and it could compute the wave field precisely, a comparison of the numerical results was undertaken with their experimental counterparts for two typical kinds of submerged bars.

Elfeki *et al.* (2007) employed a two-dimensional fully implicit finite difference model for the unsteady groundwater flow to study the influence of temporal variations in the regional hydraulic gradient and in the boundary conditions on the spreading of solute plumes in homogeneous aquifers. It was illustrated that transient flow conditions had a significant impact on contaminant transport if the amplitude and period of the oscillations were relatively large and that this tidal variation could have an effect on the spreading of solutes and on salt-water intrusion.

4.5   Three-dimensional models

Mestres *et al.* (2003) modelled the spreading of the plume induced by the freshwater discharge from the Ebro River into north-western Mediterranean coastal waters. The coastal current field was obtained with a finite difference hydrodynamic model and a Lagrangian code that solves the 3-D convection–diffusion equation and reproduces turbulent diffusion using a "random-walk" algorithm. It was found that local hydrodynamics near the river mouth, and consequently the spreading of the river plume, were highly dependent on the driving river discharge and wind field characteristics.

Pinho *et al.* (2004) delineated the hydroinformatic components and different methodologies for analysing the performance of numerical meshes, a conditioned mesh refinement procedure, a three-dimensional finite difference hydrodynamic model with an alternative technique for the external mode computation, 2-D and 3-D water quality models for coastal waters and a methodology for GIS model results integration. They suggested that the modular approach adopted in the development of this hydroinformatic environment was a very suitable and versatile methodology for decision support systems to be applied in coastal zones environment management.

Marinov *et al.* (2006) employed a 3-D hydrodynamic finite difference multi-purpose model for coastal and shelf seas (COHERENS) in Sacca di Goro, a coastal lagoon. The numerical model could be coupled to biological, resuspension and contaminant transport models and resolve mesoscale to seasonal scale processes. The investigation was mainly on the physical aspects of the modelling as an important background for the future investigation of nutrient dynamics, biogeochemical processes and contaminant transport in Sacca di Goro. An analysis was also made of the differences in temperature and salinity fields computed before and after an intervention to improve lagoon-Adriatic Sea exchange.

Abualtayef *et al.* (2008) developed and applied a three-dimensional multi-layer hydrostatic model of tidal motions in the Ariake Sea. The governing equations were solved using the fractional step method, which combined the finite difference method in the horizontal plane and the finite element method in the vertical plane. A 3-D, time-dependent, hydrostatic, tidal current model with ability to compute wetting and drying in tidal flats due to tidal motion was introduced. The model was successfully applied to Saigo fishery port and the Ariake Sea.

Carballo *et al.* (2009) investigated the residual circulation of the Ria de Muros, a large coastal embayment in north-west Spain, by employing a three-dimensional baroclinic finite difference model. Various driving forces, including tide, winds, river inflows and density forcing at the open boundary, were considered. The model was then applied to compute the residual circulation induced by the relevant agents of the Ria hydrodynamics – the tide, an upwelling-favourable wind characteristic of spring and summer, a downwelling-favourable wind typical of winter, and freshwater inflows associated with high river runoff.

4.6   A 3-D hydrodynamic and pollutant transport model

The three-dimensional hydrodynamic and pollutant transport numerical model (Chau and Jiang 2001) was evolved from the POM (Princeton Ocean

Model; Mellor 1996). The following are the key characteristics of this model:

1. The curvilinear and orthogonal coordinate system and the sigma coordinate system are employed in the horizontal and vertical directions, respectively.
2. Semi-implicit treatment is applied on the horizontal and vertical time differencing (Casulli and Cheng 1992). Whilst implicit treatment is only applied on the vertical flux term and the decay term during time integration of the governing equation, explicit treatment is applied for all remaining terms in the equation. For the horizontal time differencing of external mode, a time-splitting method is employed. The allowable time step is therefore larger than that entailed by the Courant–Friedrichs–Lewy (CFL) stability criterion $\left(dt < dx \left/ \left(\sqrt{2gh} + U_{MAX}/\sqrt{2} \right) \right. \right)$.
3. Considerations are made of the implementation of complete thermodynamics and the thermal structure of the estuary, including the density and salinity stratification as a function of temperature variation in both horizontal and vertical direction.
4. An embedded, second moment, turbulence closure sub-model is included to furnish vertical mixing coefficients.

Details on the hydrodynamic equations and the corresponding solution method can be found in Mellor (1996) whilst details on the equations for the orthogonal curvilinear transformation can be found in Chau and Jin (1995). The level of confidence, accuracy, previous calibrations and usage of the POM are detailed in Quamrul and Blumberg (1999) and Blumberg and Mellor (1987). More details of the hydrodynamic model of the PRE can be found in Chau and Jiang (2001) for this pollutant transport study. The density structure of the transporting seawater in this model is computed as a spatial and temporal function via satisfying the momentum equations, the temperature equation, and the salinity transport equations under the constraint of the prescribed boundary conditions. The major discrepancy between this model and the POM is in the second characteristic described in the above. It should be noted that in this model the horizontal time differencing is semi-implicit with the use of a time-splitting method, whilst in the POM the horizontal time differencing is entirely explicit, with the time step based on the CFL condition. The allowable time step in the POM model is therefore restricted to a value much smaller than its counterpart in this model. This characteristic will become an advantage for an application to certain domains with complex flow patterns and/or with large currents generated by tide and river discharges, such as in the application case to be described in the latter part of this chapter, namely, the Pearl River estuary.

As stated in the above, in this model, a sigma (σ) coordinate condition and an orthogonal curvilinear coordinate are employed in the vertical and horizontal directions, respectively. In the σ stretching system, σ spans the range from $\sigma = 0$ at the surface of water to $\sigma = -1$ at the bottom. It should be noted that the σ coordinate is more appropriate for simulating current flow and salinity transportation than the Z system (Leendertse *et al.* 1973) owing to its ability to furnish the same number of layers independent of water depth. Thus, it is most suitable in dealing with domains of large topographic variability. Owing to the ability of the curvilinear coordinate to replace the stagger grid in the Cartesian coordinate system and to enhance the representation of the numerical mode, it has been widely used in recent years. In fact, two kinds of curvilinear coordinate exist, namely, orthogonal and non-orthogonal. Although the orthogonal coordinate gets a simple motion equation, it is unable to generate a grid for a complex geometrical domain. Thus, for a domain with a complex boundary, a non-orthogonal coordinate system is often employed.

4.6.1 Hydrodynamic equations

The governing equations of the dynamics of a coastal cycle comprise fast-moving external gravity waves as well as slow-moving internal gravity waves. Simons (1974) presented a splitting technique to decompose the three-dimensional motion equations into a two-part sub-mode. Whilst vertical structure is an internal mode, the vertically averaged part becomes an external mode. In this way, the advantage is that it allows the computation of free surface elevation, and at the same time little sacrifice needs to be made in computational time. This is attained by isolating the solution process of the velocity transport from those of the main part three-dimensional computation processes of velocity and the thermodynamic properties (Mellor 1996). The following paragraphs show the governing equations of the two sub-modes.

By using the σ coordinate system, the internal model is a vertical structure mode depicted by the original three-dimensional equations.

Continuity equation:

$$\frac{\partial UD}{\partial x} + \frac{\partial VD}{\partial y} + \frac{\partial \omega}{\partial \sigma} + \frac{\partial \eta}{\partial t} = 0 \tag{4.1}$$

Momentum equation:

$$\frac{\partial UD}{\partial t} + \frac{\partial U^2 D}{\partial x} + \frac{\partial UVD}{\partial y} + \frac{\partial U\omega}{\partial \sigma} - fVD + gD\frac{\partial \eta}{\partial x}$$

$$+ \frac{gD^2}{\rho_0} \int_\sigma^0 \left(\frac{\partial \rho}{\partial x} - \frac{\sigma}{D}\frac{\partial D}{\partial x}\frac{\partial \rho}{\partial \sigma} \right) d\sigma = \frac{\partial}{\partial \sigma}\left(\frac{K_M}{D}\frac{\partial U}{\partial \sigma} \right) + F_x \tag{4.2}$$

$$\frac{\partial VD}{\partial t} + \frac{\partial UVD}{\partial x} + \frac{\partial V^2D}{\partial y} + \frac{\partial V\omega}{\partial \sigma} + fUD + gD\frac{\partial \eta}{\partial y}$$

$$+ \frac{gD^2}{\rho_0}\int_\sigma^0 \left(\frac{\partial \rho}{\partial y} - \frac{\sigma}{D}\frac{\partial D}{\partial y}\frac{\partial \rho}{\partial \sigma}\right)d\sigma = \frac{\partial}{\partial \sigma}\left(\frac{K_M}{D}\frac{\partial V}{\partial \sigma}\right) + F_y \qquad (4.3)$$

Temperature and salinity transport equations:

$$\frac{\partial TD}{\partial t} + \frac{\partial TUD}{\partial x} + \frac{\partial TVD}{\partial y} + \frac{\partial T\omega}{\partial \sigma} = \frac{\partial}{\partial \sigma}\left(\frac{K_H}{D}\frac{\partial T}{\partial \sigma}\right) + F_T \qquad (4.4)$$

$$\frac{\partial SD}{\partial t} + \frac{\partial SUD}{\partial x} + \frac{\partial SVD}{\partial y} + \frac{\partial S\omega}{\partial \sigma} = \frac{\partial}{\partial \sigma}\left(\frac{K_H}{D}\frac{\partial S}{\partial \sigma}\right) + F_S \qquad (4.5)$$

Turbulence energy equation:

$$\frac{\partial q^2 D}{\partial t} + \frac{\partial Uq^2 D}{\partial x} + \frac{\partial Vq^2 D}{\partial y} + \frac{\partial \omega q^2}{\partial \sigma} = \frac{\partial}{\partial \sigma}\left(\frac{K_q}{D}\frac{\partial q^2}{\partial \sigma}\right)$$

$$+ \frac{2K_M}{D}\left[\left(\frac{\partial U}{\partial \sigma}\right)^2 + \left(\frac{\partial V}{\partial \sigma}\right)^2\right] + \frac{2g}{\rho_0}K_H\frac{\partial \tilde\rho}{\partial \sigma} - \frac{2Dq^3}{B_1 l} + F_q \qquad (4.6)$$

$$\frac{\partial q^2 lD}{\partial t} + \frac{\partial Uq^2 lD}{\partial x} + \frac{\partial Vq^2 lD}{\partial y} + \frac{\partial \omega q^2 l}{\partial \sigma} = \frac{\partial}{\partial \sigma}\left(\frac{K_q}{D}\frac{\partial q^2 l}{\partial \sigma}\right)$$

$$+ E_1 l\frac{K_M}{D}\left[\left(\frac{\partial U}{\partial \sigma}\right)^2 + \left(\frac{\partial V}{\partial \sigma}\right)^2\right] + E_1 E_3 l\frac{g}{\rho_0}K_H\frac{\partial \tilde\rho}{\partial \sigma} - \widetilde{W}\frac{Dq^3}{B_1} + F_l \quad (4.7)$$

The definitions of the horizontal viscosity and diffusion terms are as follows:

$$F_x = \frac{\partial}{\partial x}(H\tau_{xx}) + \frac{\partial}{\partial y}(H\tau_{xy}) \qquad (4.8)$$

$$F_y = \frac{\partial}{\partial x}(H\tau_{xy}) + \frac{\partial}{\partial y}(H\tau_{yy}) \qquad (4.9)$$

$$F_\varphi = \frac{\partial}{\partial x}(Hq_x) + \frac{\partial}{\partial y}(Hq_y) \qquad (4.10)$$

where
$\tau_{xx} = 2A_M\frac{\partial U}{\partial x}$; $\tau_{xy} = \tau_{yx} = A_M\left(\frac{\partial U}{\partial y} + \frac{\partial V}{\partial x}\right)$; $\tau_{yy} = 2A_M\frac{\partial V}{\partial y}$; $q_x = A_H\frac{\partial \varphi}{\partial x}$; $q_y = A_H\frac{\partial \varphi}{\partial y}$;
ϕ denotes $T, S, q^2, q^2 l$; U, V, ω denotes mean water velocities in the x, y, σ directions, respectively; η is the elevation of seawater surface above the undisturbed level; f is the Coriolis parameter; $D = \eta + H$; H is the depth of the water; g is the Earth's gravitational acceleration; ρ_0 is the fluid density; ρ is the fluid density after subtraction of the horizontally averaged density;

$\tilde{\rho}$ is the buoyant fluid density; T is temperature; S is salinity; q^2 is the turbulence energy; l is the mixing length; K_M, K_H, K_q are vertical turbulent flux coefficients; A_M, A_H are horizontal turbulent coefficients; $\tilde{W}$ is the wall proximity function; and B_1, E_1, E_3 are empirical constants to be determined from laboratory experiments.

It can be observed from the above equations that a $q^2 \sim q^2 l$ turbulence model is considered and that two prognostic equations are involved, which are in essence identical to those employed in the $K \sim \varepsilon$ approach (Davies *et al.* 1995).

The external mode is written as two-dimensional dynamic equations after the depth integration of the continuity and momentum equations:

Continuity equation:

$$\frac{\partial \overline{U}D}{\partial x} + \frac{\partial \overline{V}D}{\partial y} + \frac{\partial \eta}{\partial t} = 0 \tag{4.11}$$

Momentum equation:

$$\frac{\partial \overline{U}D}{\partial t} + \frac{\partial \overline{U}^2 D}{\partial x} + \frac{\partial \overline{UV}D}{\partial y} - \tilde{F}_x - f\overline{V}D + gD\frac{\partial \eta}{\partial x}$$

$$= \langle wu(-1) \rangle - \frac{gD}{\rho_0} \int_{-1}^{0} \int_{\sigma}^{0} \left(D\frac{\partial \rho}{\partial x} - \frac{\sigma \partial D}{\partial x} \frac{\partial \rho}{\partial \sigma} \right) d\sigma \tag{4.12}$$

$$\frac{\partial \overline{V}D}{\partial t} + \frac{\partial \overline{UV}D}{\partial x} + \frac{\partial \overline{V}^2 D}{\partial y} - \tilde{F}_y + f\overline{U}D + gD\frac{\partial \eta}{\partial y}$$

$$= \langle wv(-1) \rangle - \frac{gD}{\rho_0} \int_{-1}^{0} \int_{\sigma}^{0} \left(D\frac{\partial \rho}{\partial y} - \frac{\sigma \partial D}{\partial y} \frac{\partial \rho}{\partial \sigma} \right) d\sigma \tag{4.13}$$

where $[wu(-1), wv(-1)] = -C_z \left(U^2 + V^2\right)^{1/2} (U, V)$, $\sigma \to -1$. $\overline{U}, \overline{V}$ denote the vertically integrated velocities; $\left(\overline{U}, \overline{V}\right) = \int_{-1}^{0} (U, V)d\sigma$; $\tilde{F}_x, \tilde{F}_y$ represent horizontal turbulence diffusion terms; $\langle wu(-1) \rangle$ and $\langle wv(-1) \rangle$ denote bottom stress components; and C_z represents Chezy's coefficient.

In the above momentum equations, the velocity advection, the horizontal diffusion, and density gradient represented by the second and third terms, the fourth term, and the eighth term, respectively, are integrated vertically from the corresponding terms of internal equations. Moreover, the bottom stress is generated from the velocity obtained in the internal mode. However, during the computation of the internal mode, the elevation of the water surface is acquired from the external mode. It should be noted that the truncation errors in internal and external modes are different. In other words, the vertical integrals of the internal mode velocity may differ slightly from $\overline{U}, \overline{V}$. Hence, (U, V) is adjusted to fulfil the following condition: $\int_{-1}^{0} Ud\sigma = \overline{U}$ so as to eliminate the current velocity of the internal mode.

Detailed descriptions of the internal mode can be found in Blumberg and Mellor (1980, 1987) and Mellor (1996). The method is basically semi-implicit, i.e. the treatment of all terms of momentum equations is explicit except for that of the vertical flux (the first term in the right hand side) which is in an implicit manner. Whilst all equations here are based on Cartesian coordinates, detailed descriptions of equations under orthogonal curvilinear coordinates can be found in Chau and Jin (1995). Moreover, the solution of the external mode in the original POM was entirely explicit and employed a C grid. The determination of the time step is made by taking into consideration the Courant–Friedrichs–Lewy (CFL) condition, which entails the satisfaction of the following condition: $dt < dx \big/ \left(\sqrt{2gh} + U_{\text{MAX}}/\sqrt{2} \right)$. It should be noted that, for cases with small grid sizes, the corresponding time step will be very small in order to have a stable computation, which in turns entails lengthy duration. In the development of the three-dimensional numerical model tailored for the Pearl River estuary with the smallest size of generated orthogonal curvilinear grid being 50 m, as shown in the latter part of this chapter, for implementation on a personal computer a semi-implicit method is employed in the external mode. Thus, a time-splitting alternating direction implicit scheme (ADI) on the "Arakawa C" grids (Chau and Jin 1995) as shown in Figure 4.1 is employed in this case.

x-direction:
Continuity equation:

$$\frac{\eta_{i,j}^* - \eta_{i,j}^n}{\Delta t} + \frac{\overline{U}_{i+1,j}^* \left(D_{i,j}^n + D_{i+1,j}^n \right) - \overline{U}_{i,j}^* \left(D_{i-1,j}^n + D_{i,j}^n \right)}{2\Delta x}$$
$$+ \frac{\overline{V}_{i,j+1}^n \left(D_{i,j}^n + D_{i,j+1}^n \right) - \overline{V}_{i,j}^n \left(D_{i,j-1}^n + D_{i,j}^n \right)}{2\Delta y} = 0 \qquad (4.14)$$

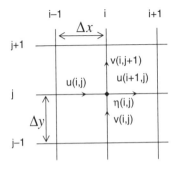

Figure 4.1 The "Arakawa C" grids

Momentum equation:

$$\frac{\overline{U}^*_{i,j} - \overline{U}^n_{i,j}}{\Delta t} + g\frac{\eta^*_{i+1,j} - \eta^*_{i,j}}{\Delta x} - f\frac{\overline{V}^n_{i-1,j} + \overline{V}^n_{i-1,j+1} + \overline{V}^n_{i,j} + \overline{V}^n_{i,j+1}}{4} = A^n \qquad (4.15)$$

y-direction:
Continuity equation:

$$\frac{\eta^{n+1}_{i,j} - \eta^*_{i,j}}{\Delta t} + \frac{\overline{V}^{n+1}_{i,j+1}\left(D^*_{i,j} + D^*_{i,j+1}\right) - \overline{V}^{n+1}_{i,j}\left(D^n_{i,j} + D^n_{i,j+1}\right)}{2\Delta y}$$

$$- \frac{\overline{V}^n_{i,j+1}\left(D^n_{i,j} + D^n_{i,j+1}\right) - \overline{V}^n_{i,j}\left(D^n_{i,j-1} + D^n_{i,j}\right)}{2\Delta y} = 0 \qquad (4.16)$$

Momentum equation:

$$\frac{\overline{V}^{n+1}_{i,j} - \overline{V}^n_{i,j}}{\Delta t} + g\frac{\eta^{n+1}_{i,j+1} - \eta^{n+1}_{i,j}}{\Delta y} + f\frac{\overline{U}^*_{i-1,j} + \overline{U}^*_{i-1,j+1} + \overline{U}^*_{i,j} + \overline{U}^*_{i,j+1}}{4} = B^n \qquad (4.17)$$

At the end of each time step the following equation has to be fulfilled:

$$\overline{U}^{n+1}_{i,j} = \overline{U}^*_{i,j} \qquad (4.18)$$

where η^* and $\overline{U}^*$ are two intermediate unknowns water stage and velocity at the first time-splitting step in the x-direction. A^n and B^n are terms acquired from the internal sub-mode. In this way, equations in each direction are written in a tri-diagonal matrix and then solved with the use of the double-sweep algorithm method (Leendertse and Crittion 1971). The solution process comprises two time-splitting steps, namely, advancing the solution from time level nt to t^* in the x-direction to determine $\overline{U}^*, \eta^*$, and then from t^* to $(n+1)t$ in the y-direction to determine $\overline{V}^{n+1}, \eta^{n+1}$. As the last step, $\overline{U}^{n+1} = \overline{U}^*$ is assumed.

It should be noted that the numerical scheme used here is semi-implicit. Hence, implicit treatments are applied on the terms of vertical diffusion in the internal mode and the elevation gradation term in the external mode. Another constraint on the time step of numerical computation entails that it cannot exceed the limits associated with the advection terms, the Coriolis term, the baroclinic pressure gradient term and horizontal diffusion term. Nevertheless, the semi-implicit numerical algorithm allows the time step to have a value much greater than that required by the CFL condition. When this model was applied in the Pearl River estuary as shown in the latter part of this chapter, the maximum time steps of the external and internal modes were 100 seconds and 60 seconds, respectively. In that application case, in order to make it simple, the same time step, namely 60 seconds, was employed in both of the two sub-modes. As a comparison, if the POM is applied in the same case study under the same grid, the maximum time steps are 6 seconds and 100 seconds for external and internal modes, respectively. Whilst this model needs a core memory of 14.8 mega words' and about 20 seconds per hour run, the POM requires 14.2 mega words' memory and 28 seconds per hour run.

It should also be highlighted that an advantage of numerical models employing sigma coordinates is the capability to address an ocean application with a large topographic variability. On the other hand, this model has the drawback of hydrostatic inconsistency due to the bottom topography effect. The horizontal density gradient along the constant σ layer is the major factor generating this inconsistency. In order to reduce this effect, several strategies adopted by Mellor *et al.* (1997) have been employed in this model. The topography is examined and the depth of water (H) is adjusted in order that it can suit the equation provided by Haney (1990) $\left| \frac{\sigma}{H} \frac{\partial H}{\partial x} \right| \delta x < \delta\sigma$. Moreover, prior to computing fluxes, a climatological density (or temperature and salinity) is subtracted. Thus, the equation of the climatological density (or temperature and salinity) fields is as follows:

$$\frac{d\rho_{\text{clim}}}{dt} = a(\rho - \rho_{\text{clim}}) \tag{4.19}$$

where a is the inverse of decay time; ρ and ρ_{clim} are density and climatological density, respectively. The value of a adopted in this Pearl River estuary model is $(1/t)$, where t is the period of major tidal constituent. A value of $t = 0.5175$ days, which equals the period of M_2, is selected in this model (Ip and Wai 1990).

4.6.2   *Pollutant transport equation*

The governing equation of pollutant transport is as follows:

$$\frac{\partial SD}{\partial t} + \frac{\partial SUD}{\partial x} + \frac{\partial SVD}{\partial y} + \frac{\partial S\omega}{\partial \sigma} = \frac{\partial}{\partial x}\left(A_S H \frac{\partial S}{\partial x}\right) + \frac{\partial}{\partial y}\left(A_S H \frac{\partial S}{\partial y}\right)$$
$$+ \frac{\partial}{\partial \sigma}\left(\frac{K_H}{D}\frac{\partial S}{\partial \sigma}\right) - K_S DS + S_s \tag{4.20}$$

where U, V, ω are the mean fluid velocities in the x, y, σ directions, respectively; S is the pollutant density as a function of x, y, σ, t, which in this case study is the density of the COD; $D = \eta + H$, where η is the elevation of the seawater surface above the mean water level, H is the mean water depth; and K_H is the vertical turbulent flux coefficient, which can be found from the second moment ($q^2 \sim q^2 \ell$) turbulence energy model (Mellor 1996). The term $q^2/2$ is the turbulent kinetic energy and ℓ is the turbulence length scale. For this numerical model, an equation is written for q^2, representing turbulent kinetic energy, and a second equation is written for $q^2 \ell$, representing turbulent dissipation. K_s is the decay rate of the pollutant and S_s is the pollutant source. A_s is the horizontal turbulence coefficient, which can be determined from the Smagorinsky formula (Oey *et al.* 1985):

$$A_s = C \Delta x \Delta y \left[\left(\frac{\partial U}{\partial x} \right)^2 + \frac{1}{2} \left(\frac{\partial V}{\partial x} + \frac{\partial U}{\partial y} \right)^2 + \left(\frac{\partial V}{\partial y} \right)^2 \right]^{\frac{1}{2}} \tag{4.21}$$

where C is a constant between 0.1 and 0.2. In this application study, a constant of 0.12 is selected, which appears to have worked well from the calibration results against standard idealized tests employed to verify the accuracy and precision of this model.

By employing the "Arakawa C" grids (Figure 4.1), the pollutant transport form are written in differencing form as follows:

$$\delta_t (SD) + \delta_x \left(\overline{S}^x \overline{D}^x U \right) + \delta_y \left(\overline{S}^y \overline{D}^y V \right) + \delta_\sigma \left(\overline{S}^\sigma \omega \right)$$

$$= \delta_x \left(\overline{H}^x \overline{A_s}^x \delta_x S \right) + \delta_y \left(\overline{H}^y \overline{A_s}^y \delta_y S \right) + \delta_\sigma \left(\frac{\overline{K_H}^\sigma \delta_\sigma S_+}{D} \right) - K_S DS_+ + S_s \tag{4.22}$$

In equation (4.22), for any parameter F as a function of x, y, σ, t, i.e., $F = F(x, y, \sigma, t)$, the following equations can be written:

$$\delta_t F = \frac{1}{2\Delta t} \left[F(x, y, \sigma, t + \Delta t) - F(x, y, \sigma, t - \Delta t) \right] \tag{4.23}$$

$$\overline{F}^x = \frac{1}{2} \left[F \left(x + \frac{\Delta x}{2}, y, \sigma, t \right) + F \left(x - \frac{\Delta x}{2}, y, \sigma, t \right) \right] \tag{4.24}$$

$$\delta_x F = \frac{1}{\Delta x} \left[F \left(x + \frac{\Delta x}{2}, y, \sigma, t \right) - F \left(x - \frac{\Delta x}{2}, y, \sigma, t \right) \right] \tag{4.25}$$

$$S_+ = S(x, y, \sigma, t + \Delta t) \tag{4.26}$$

It should be noted that, in the differencing equation (4.22), all components can be determined from the previous time step of the hydrodynamic model, except for the following three unknowns: $S(x, y, \sigma, t + \Delta t)$, $S(x, y, \sigma + \Delta \sigma, t + \Delta t)$, and $S(x, y, \sigma - \Delta \sigma, t + \Delta t)$ in the first term of the left-hand side and the third and fourth term of the right-hand side of the equation, respectively. Equation (4.22) is therefore re-written as follows:

$$AS(x, y, \sigma - \Delta \sigma, t + \Delta t) + BS(x, y, \sigma, t + \Delta t)$$

$$+ CS(x, y, \sigma + \Delta \sigma, t + \Delta t) = D \tag{4.27}$$

where A, B, C, D are known coefficients. It can be observed that equation (4.27) is a tri-diagonal matrix in the vertical direction which can be solved with the technique detailed in Richtmyer and Morton (1967).

The validation of this model has been undertaken by performing several tests involving idealized geometries and forcing functions, the same as those previously applied to the POM during its development stage. They include not only simple tests on checking the ability of the model to conserve its various constituents, but also more rigorous tests involving both barotropic and baroclinic responses of an idealized coastal basin with or without topography to evolve different large-scale oceanographic phenomena (Blumberg and Mellor 1987). It was shown that the model reproduced the expected physics and produces identical results to those from the well-tested POM (Chau and Jiang 2001). These results furnished a high degree of confidence that the numerical accuracy of the scheme is consistently high and that the level of numerical diffusion is not larger than the physical diffusion computed in the model.

4.7 Advantages and disadvantages

The relative merits of the numerical schemes can be compared on the basis of the computational stability, convergence, accuracy, and efficiency. The convergence and stability of a scheme depend strongly upon the finite difference formulation used and upon the initial and boundary conditions.

It is noted that the difference quotients used are, in fact, a truncated Taylor's series. The degree of the approximation represented by the finite difference analogues is called the "truncation error" or "order of approximation". The truncation error may be found by expanding the various terms in the differential equations into a Taylor's series. But even when the solution of the difference equations converges to a solution of the differential equations, it is not necessary that the numerical solution of these difference equations approaches the required solution. It is possible that when the computation progresses, waves are generated by the computational procedure that overshadow the actual solution, and the solution is unstable. These spurious solutions are caused by the inevitable rounding errors, which influence the required solution of the finite difference equations.

For a given problem the rate of growth of the instability depends on the choice of Δx and Δt. The instability may be controlled and marked by the natural damping supplied by the friction term. If natural attenuation and dispersion (due, for example, to the friction term in the momentum equation) are much stronger than numerical damping and dispersion, there is no need to worry too much about them. Such is the case for many practical applications. However, numerical damping can be a nuisance when the waves are of relatively small length and friction is negligible – such as in the case of steep front waves in canals and some tidal rivers.

The concept of convergence assumes that a sequence of computations with an increasingly finer mesh tends toward the exact solution. By so defining it may loosely be considered as an indication of accuracy. Thus the convergence qualities of a scheme may be of decisive importance as to a modelling system's usefulness. In practice, the convergence can be achieved by using reasonably small Δx and Δt. If convergence is ensured, its rapidity depends on the order of approximation. But the order of approximation of derivatives by finite differences may well be a meaningless notion for real computational grids because the time and space intervals Δt, Δx and the derivatives of variables may well not be small.

Because all numerical methods are approximations, the question of numerical accuracy must be considered. One measure of accuracy is the degree of difference between observed real-life data (such as measured hydrographs) and computed results. This criterion is obviously most important but it is practically impossible to formulate. There might be several reasons for discrepancy between a mathematical model and the prototype, such as inaccurate simplifications and approximations in the basic equations failing to simulate the complexity of the prototype, insufficiently accurate measuring techniques, insufficient data, phenomena which are not taken into account and poor schematization of topographic features.

An additional factor which should be considered in comparing different computational methods is the computer time required and the difficulties in programming. No generally agreed yardsticks have been established concerning this aspect. The precise comparison depends on the particular scheme and equations used and the logic in programming. For explicit schemes the computational effort depends mainly on the time increment Δt used, which is determined by some stability criteria for given Δx. However, the maximum values of Δx that can be used are often limited by the channel geometry and accuracy criterion; this can sometimes result in excessively small Δt and long computer time. Generally speaking, compared to an implicit scheme, the computational effort per time step is less for an explicit scheme; however, the number of time steps may be unreasonably large (due to stability constraints).

The explicit method appears to be superior to both the implicit method and the methods of characteristics in terms of simplicity of the programming. However, all explicit finite difference schemes, when applied to the hyperbolic flow equation, are conditionally stable. The allowable time step is thus limited by the grid size used and the numerical consideration, and not necessarily by the time-scale of the physical phenomena under consideration.

4.8 Applications and case studies

A prototype application of the numerical model is on the Pearl River estuary, which includes four outlets of the Pearl River system and the main part of Hong Kong seawaters. So far, this is probably the first application of this type of model (3-D and baroclinic) in the PRDR. It is certain that the model, as a decision-supporting tool, has a significant value from an engineering point of view for exploring the dynamics and circulation of the PRDR. The computation is then calibrated and verified with field observation data.

4.8.1 Description of the Pearl River estuary

The Pearl River estuary (PRE) has been considered one of the most important zones in South China. With the rapidly developing economy in this area, the environment has been gradually deteriorating, thus in turns attracting more and more attention on this area of estuary. Since the coupled hydrodynamic and pollutant transport numerical model is an efficient tool for environmental impact assessment and feasibility study of projects, the model has been applied to this estuary with calibration.

The study area (Figure 4.2) is a delta estuary comprising four main river outlets (namely, Hu men, Jiao men, Hongqi men, Heng men) in the

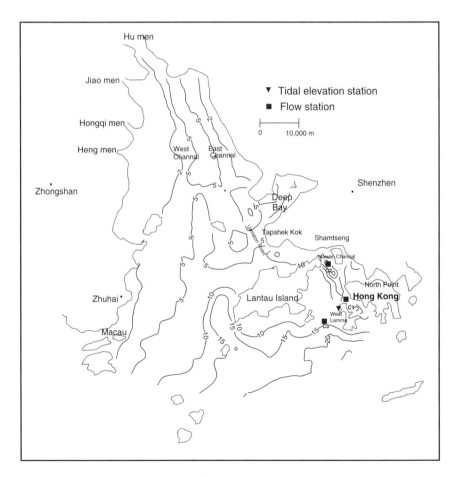

Figure 4.2 Topography of Pearl River estuary

north-west of the PRE and the Shenzhen River outlet into the Deep Bay. Whilst the water depth at the open end of the estuary ranges between 20 m and 28 m, it becomes shallower toward the inner bay. Overall, the mean depth of the estuary is 7 m. The average net discharges of the four outlets for different seasons are listed in Table 4.1 for years from 1985 to 1995 (Pang and Li 1998). It can be observed that the variations in discharge are large in different years. The predominant tide in the PRE is a semi-diurnal and irregular tide with a mean tidal range of 1.0 m or so. Whilst at the entrance to the estuary, the mean tidal range is 0.85–0.9 m, its value increases into the inner estuary to 1.6 m at the Hu men River mouth (Kot and Hu 1995). During the wet season, which occurs from May to September, the runoff of the rivers is strongest, to become the predominant hydrodynamic forcing in the Pearl River estuary. During the dry season, which is between December and March, the tidal current is the major forcing function (Lu 1997).

Figures 4.3 and 4.4 show the horizontal grid of the orthogonal curvilinear coordinate system and the corresponding transformed grid, respectively. There are 3,400 grids in each of the six vertical layers, and each layer has the same $\delta\sigma$ with a value of 1/6. The number of layer and grid points is selected in order to attain reasonable accuracy in both horizontal and vertical discretizations, but at the same time not to affect adversely the computational efficiency.

4.8.2 Boundary and initial conditions

The tide elevation is the model forcing at the two open boundaries, namely, the southern boundary (South China Sea) and the eastern boundary (Lei Yu Mun). Its value can be interpolated from the observed data at two tidal stations, namely, Macau and North Point, by using the tidal wave propagating speed $\sqrt{gh}$ (Huang and Lu 1995). The current velocities of external and internal modes at the open boundaries are determined from the radiation condition; for example $\frac{\partial v}{\partial t} - c_i \frac{\partial v}{\partial y} = 0, c_i = \sqrt{H/H_{\max}}$ at the open boundary in the South China Sea. The salinity, temperature and turbulence kinetic and turbulence dissipation at these open boundary conditions are determined as follows:

Ebb time:

$$\frac{\partial A}{\partial t} + U \frac{\partial A}{\partial x} = 0 \tag{4.28}$$

Flood time:

$$A = A_{\text{set}}(t, \sigma) \tag{4.29}$$

Table 4.1 The discharge of four east outlets in different seasons (wet season months: May–September; mean season months: April, October and November; dry season months: December–March) (unit: 10^8 m^3/ season)

	Hu men			Jiao men			Hongqi men			Heng men		
	Wet	Mean	Dry	Wet	Mean	Dry	Wet	Mean	Dry	Wet	Mean	Dry
1985	303.5	158.7	108.1	290.2	146.4	93.5	112.6	51.6	27.6	190.4	90.6	59.8
1986	341.4	126.2	58.9	326.4	116.4	50.9	126.7	41.0	15.0	214.2	72.1	32.6
1987	284.4	149.3	66.2	271.9	137.8	57.3	105.5	48.5	16.9	178.4	85.3	36.6
1988	294.5	123.6	72.4	281.6	114.0	62.6	109.3	40.2	18.4	184.8	70.6	40.0
1989	219.1	97.4	68.0	209.5	89.8	58.7	81.3	31.7	17.3	137.5	55.6	37.6
1990	260.9	187.5	96.2	249.5	172.9	83.2	96.8	60.9	24.5	163.7	107.1	53.2
1991	242.6	79.3	63.6	232.0	73.2	55.0	90.0	25.8	16.2	152.2	45.3	35.2
1992	289.7	133.2	109.0	277.0	122.9	94.2	107.5	43.3	27.8	181.8	76.1	60.3
1993	397.2	131.9	61.5	379.8	121.7	53.2	147.4	42.9	15.7	249.2	75.3	34.0
1994	500.9	171.5	84.4	478.9	158.2	73.0	185.8	55.7	21.5	314.2	97.9	46.7
1995	307.8	188.1	104.9	294.3	173.5	90.7	114.2	61.1	26.8	193.1	107.4	58.0
Average	312.9	140.6	81.2	299.2	129.7	70.2	116.1	45.7	20.7	196.3	80.3	44.9

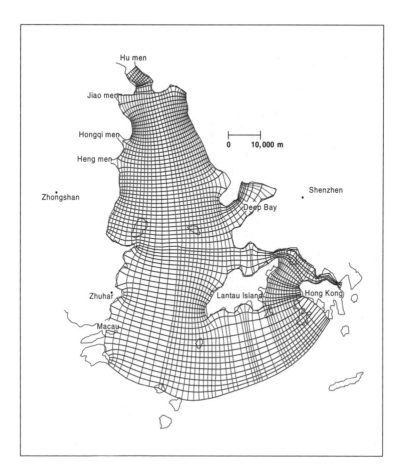

Figure 4.3 The orthogonal curvilinear grid

where A represents the salinity, temperature, turbulence kinetic and turbulence dissipation. During the ebb process, at open boundaries, A are computed using the "upwind" differenced advection equation. During the flood stage, A is linearly interpolated from its value at the end of the ebb process to a constant A according to the depth and observed data. The open boundary conditions of four outlets in the north-western part of the study area are determined by water discharges.

Simple treatment is often applied to the open boundary condition for pollutant transport (Leendertse and Crittion 1971). For example, employing grids near the eastern open boundary at Lei Yu Mun, the equations are as follows:

$$P_{i,j}^{n+1} = P_{\text{set}} \quad U_{i-\frac{1}{2},j}^{n+1} < 0 \tag{4.30}$$

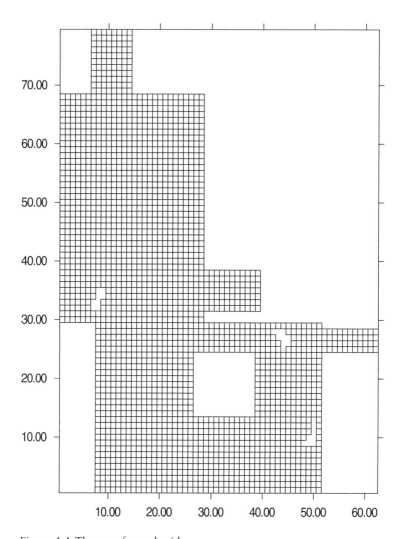

Figure 4.4 The transformed grid

$$\frac{dP}{dt} = 0 \quad \text{that is} \quad \frac{P_{i,j}^{n+1} - P_{i,j}^{n-1}}{2\Delta t} + U_{i-\frac{1}{2},j}^{n+1} \frac{P_{i,j}^{n} - P_{i-1,j}^{n}}{\Delta x} = 0 \ U_{i-\frac{1}{2},j}^{n+1} > 0 \quad (4.31)$$

where P_{set} denotes the prescribed along-boundary component of pollutant density. Equation (4.30) is for flood tide while equation (4.31) applies during the ebb tide condition. The value P_{set} can be applied if it is known. However, it is often unknown and an assumed value has to be used according to the best available data. In this way, the open boundary condition stated above is reasonable if either the boundary condition is known or the level of water exchange and/or flushing outside the model domain is so strong that no internal pollutant build-up occurs at the open boundary. It is noted that no COD data are available in the PRE, and the exchange capacity at the entrance of the PRE is not strong enough that the $P_{set} = 0$ approach can be employed. For the flood condition, a simple and efficient open boundary condition for pollutant transport is employed:

$$\frac{P_{i,j}^{n+1} - P_{i,j}^{n-1}}{2\Delta t} + U_{i-\frac{1}{2},j}^{n+1} \frac{(a-1)P_{i-1,j}^n}{\Delta x} = 0 \quad U_{i-\frac{1}{2},j}^{n+1} < 0 \tag{4.32}$$

Under this approach, the same equation (4.31) is used at ebb tide, but at flood tide equation (4.32) is used to replace equation (4.30). Equation (4.31) denotes that there is no spatial gradient in concentration. Equation (4.32) denotes the proportion of pollutant concentration brought back by the flood tide. The constant coefficient, a, in equation (4.32) ranges from 0 to 1, which is dependent on the level of water exchange/flushing outside the model domain. If the level of flushing is strong, the value of a becomes small; otherwise, the value becomes larger and approaches 1. Figure 4.5 shows the COD concentration at the boundary corresponding to different a values when stable computation is attained. Moreover, a lower limit is imposed on the boundary condition in equation (4.32), namely, $P_{i,j} = 0$ if $P_{i,j} < 0$. As shown in Figure 4.5, this condition happens in some tidal periods at small a values. In this case study, $a = 0.9$ is selected, which appears to work well according to the internal COD calibration. According to the initial background COD condition, if the COD concentration at the boundary is less than 1.8 mg/L, it is prescribed to become 1.8 mg/L.

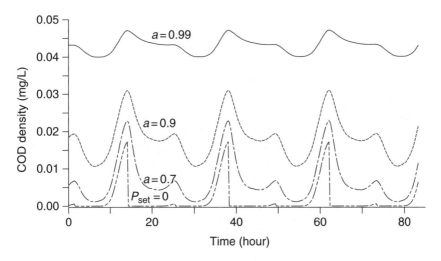

Figure 4.5 Time variation of COD density at the open boundary at different a values

In the motion equations, the solution of convection and diffusion terms requires the value of velocity at the outer boundary. Two kinds of closed boundary conditions have been attempted in this model, namely, no-slip condition for $\partial u/\partial y$ assuming $u = 0$, and free slip condition with $\partial u/\partial y = 0$. In this model, these two methods are weighted together to obtain the semi-slip boundary condition: $\partial u/\partial y = \beta u/\Delta y, 0 \leq \beta \leq 1$. Sensitivity tests with different β values have been performed, which revealed little change in results.

The initial pollutant density of COD in the model domain was assumed to have a constant background value of 1.8 mg/L. After a number of computational tidal periods (in this model 100 tidal periods, which is about 50 days), a steady state concentration gradient is attained.

4.8.3   Calibrations

A hydrological survey was conducted by the Hong Kong Civil Engineering Department Port Development Division for one whole year. The field data obtained at three tidal elevation stations and three tidal current stations were employed to calibrate this model. Figure 4.2 shows the location of these stations. The observed data of tidal elevation on two tidal stations, Macau and North Point, are employed as boundary conditions. Simulation of one month's hydrodynamics of the Pearl River estuary is performed.

For calibration of the computer simulations, the results of tidal elevation, flow velocity and flow direction have been verified with the corresponding observed data. Owing to the lack of detailed salinity data, comparison of salinity results is only made with previous study by others. The comparison of simulated and observed tidal elevations for the whole month at West Lamma is shown in Figure 4.6. However, a typical day of comparison can demonstrate better the slight differences between the computed and simulated results more effectively. Figure 4.7 shows the simulated and observed results of tidal elevation at three different elevation stations for

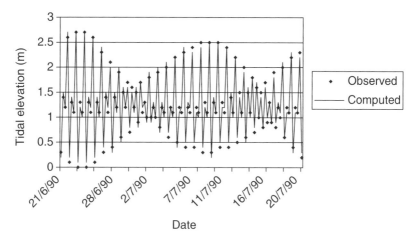

Figure 4.6 Comparison of computed and observed tidal elevations at West Lamma

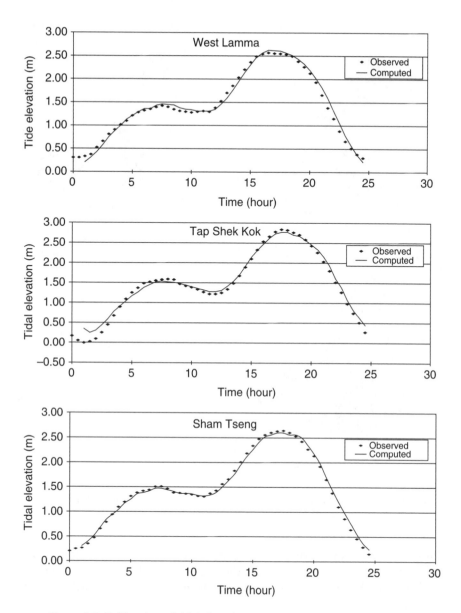

Figure 4.7 Calibration of tidal elevation

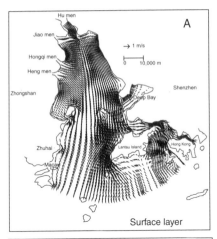
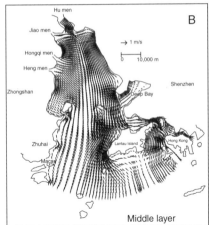
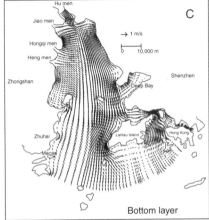

Figure 4.8 Flow field during ebb tide in mean season at different layers

24 hours, which shows that the computational results concur well with the measured data. It is apparent that the tidal amplitude at an inner station such as Tap Shek Kok is larger than that at the outer stations such as West Lamma.

Figure 4.8 shows the horizontal tidal current pattern in the Pearl River estuary during an ebb tide in mean season (representing April, October and November) at the surface, middle, and bottom layers, respectively. It can be observed that the current velocities at the sea surface are slightly larger than that in the bottom. The flow directions at the surface layer and bottom layer may be slightly non-uniform, especially under low to medium current. The maximum flow velocity of 2.5 m/s or so occurs at narrow channels in Ma

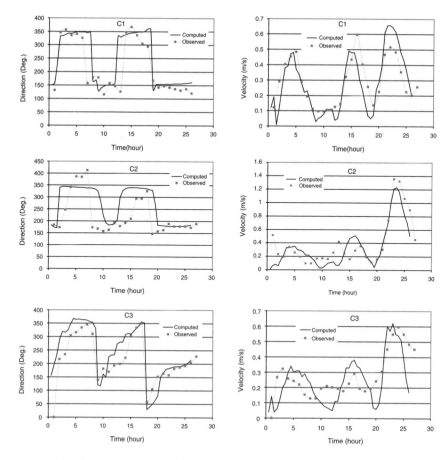

Figure 4.9 Calibration of tidal flow direction and velocity for one day at three stations

Wan Channel and Kap Shui Mun with water depth generally deeper than 20 m. The current speed in north-west Lantau Island water areas, Urmston Road and West East Channel are higher than other locations. This can be justified because the bathymetry of these locations causes lateral contraction in the flow channel and in turns generates faster flow. The simulated current pattern in general agrees well with the field data (Kot and Hu 1995).

Figure 4.9 shows the results comparisons of direction and magnitude of depth-averaged velocities at three tidal stations between computed and observed data from 15:30 21st to 22nd June. The root-mean-square (rms) errors of the computed tidal level, flow direction and velocity for the one month comparisons are 0.14 m, 17 degrees, and 0.07 m/s, respectively. The computed flow direction and velocity agree well with the field data, which serves as another proof that the model can simulate well the field data. It

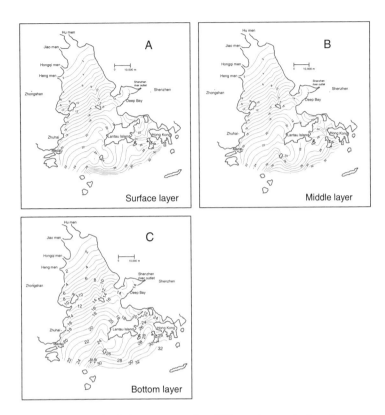

Figure 4.10 Computed salinity contours for different layers during the wet season

can also be noted that the start-up of the simulation needs about 2 to 3 hours prior to produce the convergence to the accurate solution compared with the field data.

Simulations of the distribution of salinity in the wet and dry seasons are also made. The measured data including tidal level and salinity at open boundaries between June and July are employed as boundary conditions during the wet season. Owing to a lack of data during the winter season, the same tidal levels are employed to simulate the salinity distribution in this dry period. The salinity values at the open boundaries are determined from the dry season salinity horizontal and vertical patterns furnished in Kot and Hu (1995) and Broom and Ng (1995). The discharges of the four rivers during wet and dry seasons are as shown in Table 4.1. Figure 4.10 shows the distribution of salinity during ebb tide in three layers during the wet season. It can be observed that, during the wet season, a sharp change of salinity is a common phenomenon in the Pearl River estuary. It is also

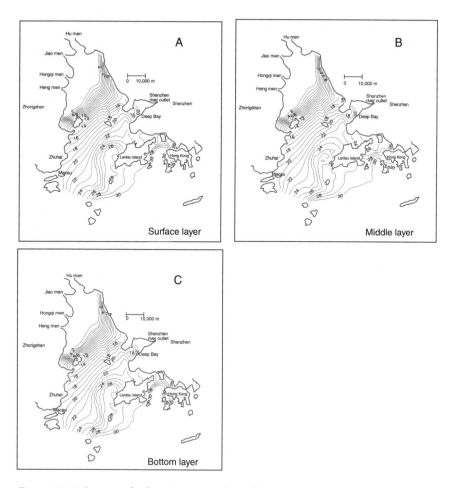

Figure 4.11 Computed salinity contours for different layers during the dry season

shown that the stratification is discernible only in the outer bay, not in the inner bay. This is possibly because the water depth is less and the turbulence is stronger there. Figure 4.11 shows the distribution of salinity during ebb tide in three layers during the dry season. It can be noted that the sharp gradient of salinity or stratification is less notable and that the seawater intrusion is located even farther at the inner delta area. Thus, in the inner delta, although the gradient of salinity is smaller, the averaged salinity is higher than that in the summer season. It is found that the pattern of the distribution of salinity is similar to that described by Kot and Hu (1995).

Table 4.2 Summary statistics of wastewater flow and COD load from Guangdong Province (unit: million ton)

Source	1990	1991	1992	1993	1994	1995	1996
Domestic	1110.12	1058.47	1310.49	1492.91	1981.31	2123.69	2122.22
Industrial	1402.5	1392.24	1419.39	1397.62	1315.31	1609.79	1508.74
Total wastewater	2512.62	2510.9	2801.03	2959.9	3372.07	3816.57	3714.05
COD	0.69	0.72	0.78	0.8	0.94	1.12	1.06

Source: Guangdong Province Yearbook Editorial Committee (1996).

A key objective of the application of the numerical model is to examine pollutant transport in the PRE and to study the distribution of pollutants during different seasons. Figure 4.2 shows four main river outlets discharging pollutants to the PRE located in the north-west PRDR and another from the Shenzhen River near Deep Bay. Since no direct COD data from the different river outlets are available, the loadings of COD at different river outlets in this model are determined in the following paragraph.

Table 4.2 lists the domestic and industrial wastewater flow from the Guangdong Province based on the Guangdong Yearbook Editorial Committee (1996). An empirical relationship between the COD loading rate and the domestic and industrial wastewater flow rates can be found accordingly:

$$W_{COD} = 0.00027Q_d + 0.000305Q_i \tag{4.33}$$

where W_{COD} is the COD loading rate, Q_i is the flow rate of the industrial wastewater, and Q_d is the flow rate of the domestic wastewater. With the use of this empirical equation, the COD loading rates from eight cities around the PRDR can be determined from the pertinent wastewater flows. Table 4.3 shows the computed COD loading rates from eight cities around the PRDR for 1996. Since the Pearl River comprises a river network system, the COD loading data at different main river outlets are approximated from the COD loading of the eight cities shown in Table 4.3. Whilst Table 4.4 lists the COD loading at the five main river outlets, the COD loading data in the Hong Kong Special Administration Region (HKSAR) are estimated from the Hong Kong strategic sewage disposal plan (Sin *et al.* 1995). All COD loadings are assumed to be point sources discharged continuously at specific locations. Another assumption in this model is that there is no sink term representing loss rate of COD.

Table 4.3 The quantity of COD discharges from different cities in 1996 (unit: 10,000 ton)

Source in 1996	Guangzhou	Shenzhen	Zhuhai	Huizhou	Dongguan	Zhongshang	Jiangmen	Foshan
Domestic	78990	11645	7113	16701	4170	5866	6895	22627
Industrial	32325	5501	3054	1926	10778	9645	11126	16911
Total wastewater	111315	17146	10167	18627	14948	15511	18021	39538
COD	31.2	4.8	2.9	5.1	4.4	4.5	5.3	11.3

Source: Guangdong Province Yearbook Editorial Committee (1996).

Table 4.4 Net water discharge and COD load of different river outlets during different seasons

	Hu men	Jiao men	Hongqi men	Heng men	Shengzhen
Wet(10^8 m³/day)	2.09	1.99	0.77	1.31	0.06
Mean(10^8 m³/day)	1.56	1.44	0.51	0.89	0.06
Dry(10^8 m³/day)	0.68	0.59	0.17	0.37	0.06
COD(kg/day)	115068	309000	70285	12300	89877

4.8.4   Simulated results

In this model, the assumed COD loadings from the PRDR and the HKSAR in different seasons have been employed to simulate the distribution of COD generated by these pollutant sources together with the background sources. Moreover, the evaluation of the effect of the pollutant sources from the PRDR on Hong Kong seawaters is made via a sensitivity analysis.

As a result, the computed salinity contours indicate very slight vertical density stratification and strong vertical mixing. The difference of COD concentrations between the top and bottom levels is very small, which in turns justifies the use of a vertically averaged value. Figure 4.12 shows the average distribution of COD in the study area during different seasons at ebb tide. It can be observed that the COD concentration in the western PRE depends on the season. During the wet season, the COD concentration in north-western PRE is lower. The reason may be the higher dilution associated with the larger average discharge flow. On the other hand, the concentration in south-western PRE during the wet season is higher than that in the dry season. This may be because of the higher conveyance in the wet season. However, the variation of COD concentration with season in the eastern PRE is small. This may be explained by the fact that a tidal current dominates the hydrodynamic forcing and that the boundary condition remain unchanged at different seasons.

Model calibration was performed at two sections, namely, the longitudinal section A_1–A_8 and the latitudinal section C_1–C_7. The field observation data was monitored by Wen *et al.* (1994). Figure 4.13 displays both the simulated and measured COD data for these two sections during different seasons. It can be found that the simulated results tend to overestimate the actual measured data in most cases, except at the western side of the estuary. Nevertheless, given the condition that there is a lack of actual COD loading data from different river outlets, the accuracy of the results by this model can be considered quite satisfactory. For future works, an improvement of the accuracy of the results can be accomplished by the collection of the actual COD loadings from all river outlets.

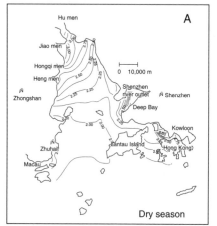

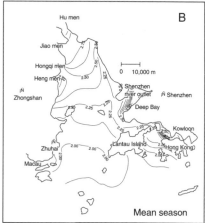

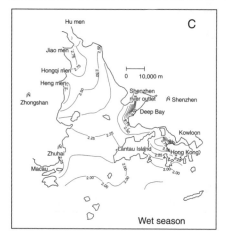

Figure 4.12 Mean COD distribution (in mg/L) at various seasons

A model sensitivity analysis is also undertaken to estimate the impact of sewage pollutants, such as COD, discharged from the PRDR on water quality in the seawaters of HKSAR. This was performed by imposing the COD sewage loading from the PRDR and at the same time assuming the background COD value to be 0 mg/L. Figure 4.14 shows the results from the wet season whose pollutant transport has the highest value amongst different times. It is found that the impact of COD from the five river outlets in the PRDR can be up to the north-western part of Lantau Island whilst its impact on other areas of Hong Kong seawaters is less significant. The increase in COD concentration generated by the loadings of the five outlets is greater than 0.25 mg/L near Lantau Island. During the wet season,

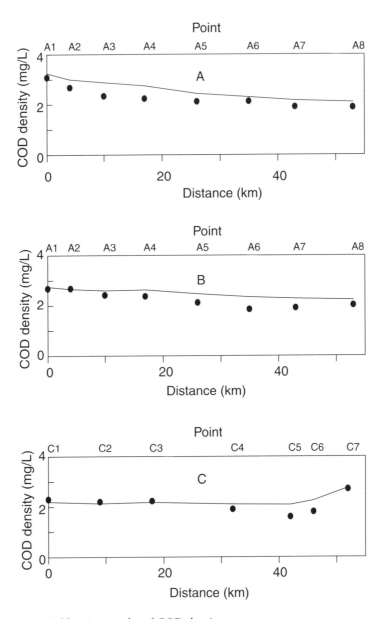

Figure 4.13 Calibration results of COD density
• monitored data, — computed result (A – along the longitudinal section during dry season; B – along the longitudinal section during wet season; C – along the latitudinal section during mean season)

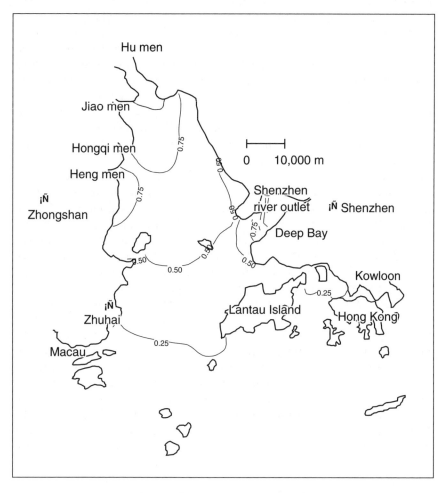

Figure 4.14 Increment of COD density (mg/L) resulting from sewage loading from the PRDR during wet season

the COD is transported over an extensive area with the assistance of the highest flow value, and in turns influences the water quality of Hong Kong seawaters to the largest extent.

4.9   Conclusions

In this chapter, a three-dimensional coupled hydrodynamic and pollutant transport model has been formulated, verified and applied to the Pearl River estuary, which is one of the most quickly developing regions in China. Hong Kong and Macau are at its entrance. In this model, the external and internal gravity waves are split into a two-dimensional mode and a three-dimensional mode, respectively. In the external mode, implicit treatment is applied to the tidal elevation gradient term whilst the other terms are integrated or acquired from the internal mode. In the internal mode, implicit treatment is applied to the vertical flux term of momentum equation whilst tidal elevation values are obtained directly from the external modes. A consistent time step is selected in both internal and external mode. In this way, the time step is larger than that allowed in the POM model and thus less computational time is imposed to attain a stable computation.

The model is applied to the Pearl River estuary taking into consideration the stratification resulting from both salinity and temperature. The simulated results of tidal elevation, both direction and magnitude of current velocity, are compared with real observation data. Results indicate good agreement between the two sets of data. The distributions of salinity in different depth layers during both the wet and dry seasons are shown. It demonstrates the development of a complicated and efficient three-dimensional model, which works well in a typical estuary, namely, the Pearl River estuary. It is shown that a three-dimensional, numerical model based on an orthogonal curvilinear grid system in the horizontal direction and a sigma coordinate system in the vertical direction for predicting water quality constituents is developed. A simple and efficient open boundary condition for pollutant transport is employed in this model. The model is employed to simulate the distribution of COD in the PRE and to assess the transboundary pollution between Guangdong and Hong Kong. Although the pollutant load data at the five main river outlets to the PRE are not available directly, the COD loading rates are determined based on the available pollutant sources. Results indicate that the pollutants from the PRDR have significant effects on the Hong Kong seawaters, in particular during the wet season, when the water discharge from the upstream of the estuary is large.

5 Finite element methods

5.1 Introduction

This chapter describes finite element methods often employed in coastal hydraulics in general and highlights a characteristic-based Galerkin method suited for advection-dominated problems (Chau *et al.* 1991; Chau 1992b). A two-step algorithm is also presented which significantly reduces the computational cost. It is shown for the 1-D scalar advection equation that the time-discretized equation can also be obtained by following a characteristics approach. The governing hydrodynamic and mass transport equations are written in conservative form, in order to exploit the full power for the numerical technique. An error analysis of the scheme using linear elements for spatial discretization is given for the one-dimensional advection-dominated equation. Remarks on advantages and disadvantages of the characteristic-Galerkin method are then made. Two applications of this robust finite element model in engineering practice have been attempted. The model is first applied to compute tidal current and advective mass transport in Tolo Harbour, Hong Kong. A scalar pollutant is released at the Yung Shue Au fish culture zone in Three Fathoms Cove. The flushing rates of semi-enclosed bays in the harbour are determined numerically via a solution of the full equations with a realistic tidal boundary condition and over half spring-neap cycle. The model is also applied to study the effect of proposed massive reclamation under the Hong Kong Port and Airport Development Strategy (PADS) project on tidal current in Victoria Harbour, Hong Kong, which is a more difficult task since there are two open boundaries.

5.2 Basic philosophy

The fundamental theoretical concept of finite element methods involves the formulating a variational boundary value problem statement of system energy and generating a discretized approximation of its extremum employing the pertinent concepts and procedures. It was originally developed by structural engineers to analyse large aircraft systems (Turner *et al.* 1956). Widespread applications of finite element methods to various non-structural problems, including fluid mechanics, began in the 1970s (Oden 1972; Baker 1973; Chung and Chiou 1976). To this end, the concept of force balance in structural mechanics was replaced by a robust theoretical analysis founded in the equivalent variational Rayleigh–Ritz methods (Rayleigh 1877; Ritz 1909).

It is noted that the direct application of these classical theoretical concepts for cases in fluid mechanics is not feasible. Yet, for most of these cases, the algorithm constructions can be regarded as specific criteria within a weighted residual framework. In other words, instead of the error in the approximate satisfaction of the conservation equations, the integral with respect to selected weights is set to zero (Finlayson 1972). Whilst in classical mechanics, one often directly applies the momentum equations (Newton's Second Law), finite element principles have evolved employing the alternative approach of energy minimization. In fact, the collocation method within weighted residuals produces the same effect as quotients in finite difference methods. This provides a link between finite element variational principles and finite difference handling of partial differential equations in fluid mechanics fields. Currently, the Galerkin criteria weighted-residuals formulation is often employed as the most representative extension from the classical concepts. It generalizes the weights to be functions and defines them as identical to the approximation functions for the conservation variables (Galerkin 1915). In fact, the least squares method is yielded when the weights are defined to be the differential approximation error.

In the following few sections, some aspects of finite element methods as applied to progressively more complete differential equation descriptions in coastal hydraulics are delineated. For all cases, the basic Navier–Stokes equation system governs completely general flow fields in three-dimensional configurations. In each case, the system is simplified by enforcing different assumptions with different degrees of restrictions on certain physical and/or geometric aspects.

5.3 One-dimensional models

In fact, one-dimensional models are not often used. The advances of computing technology in these few decades mean that the computing effort is no longer a controlling factor. In general, higher accuracy attached to higher dimensional models will be the more deciding factor in the choice of numerical model.

Hagen *et al.* (2001) proposed a localized truncation error analysis (LTEA) as a means to efficiently generate meshes that incorporate estimates of flow variables and their derivatives. Three different LTEA-based finite element grid generation methodologies were examined and compared with two common algorithms: the wavelength to Delta x ratio criterion and the topographical length scale criterion. Errors were compared on a per node basis. It was demonstrated that solutions based on LTEA meshes were, in general, more accurate in both local and global terms, and were more efficient. Analyses and results from this 1-D study could be considered to have laid the groundwork for developing an efficient mesh-generating algorithm suitable for higher dimensional models.

Muttin (2008) presented two numerical finite element models, namely, a 1-D model (FORBAR) and a 3-D model (SIMBAR), in simulating an oil-spill boom, a long floating structure used to deviate or to stop floating pollution during an oil-spill crisis. The objectives of these models were mainly to furnish as much mechanical information as required to optimize the contingency system and to formulate appropriate emergency action plans in time, i.e. deviative strategy or stopping strategy. The models were applied in the case of the Elorn River in France. Comparisons of the results were made with the observed field data as the benchmarking yardstick.

5.4 Two-dimensional models

Whilst 2-D depth-averaged models are often employed in most cases in coastal engineering, 2-D laterally averaged models will be used when the relationship between the pertinent parameters and the water depth is of great concern.

5.4.1 2-D depth-integrated models

Fernandes *et al.* (2002) used a two-dimensional depth-averaged finite element flow model (TELEMAC-2D) to model the hydrodynamics of the Patos Lagoon during the 1998 El Niño event. The model was initially calibrated against measurements taken over a period of time. Then model validation was carried out by comparing measurements and predictions for a reference station in the estuarine area for a different period. Results indicated that velocities in the lagoon and estuary during the extreme conditions observed in the El Niño period were much stronger from those during the normal periods.

Keen *et al.* (2004) applied a 2-D finite element hydrodynamic model to simulate the oceanographic and sedimentological processes that produced the event beds in Mississippi Sound and the inner shelf of the north-east Gulf of Mexico. The simulation was employed to validate whether or not the simulated cores were consistent with the observed stratigraphy and geochronology and whether or not the event beds were probably produced by an unnamed hurricane in 1947 and by Hurricane Camille in 1969.

Sentchev *et al.* (2006) presented a two-dimensional (2-D) finite element spectral in time model to describe the Stokes drift of the major tidal constituents in the English Channel, assimilating high frequency radar surface velocities and coastal tidal gauge data. The assimilated sea surface height and depth-averaged velocities, based on six major tidal constituents, were employed for the mapping of the residual transport through the Channel, tidal dissipation, and for estimation of the energy flux. Error charts for the sea surface elevation demonstrated that some local coastal areas within the channel might need better coverage by observations.

Mattocks and Forbes (2008) developed a real-time, event-triggered storm surge prediction system for the State of North Carolina to assist government officials with evacuation planning, decision-making and resource deployment during tropical storm landfall and flood inundation events. The system was accomplished through a two-dimensional, depth-integrated version of the ADCIRC (Advanced Circulation) coastal ocean model with winds from a synthetic asymmetric gradient wind vortex.

5.4.2   2-D lateral-integrated models

Sakr (1999) developed a two-dimensional finite element model, 2D-VDTRAN, to simulate density-dependent solute transport in solving the problem of seawater intrusion for the case of a confined coastal aquifer in which there is steady seaward flow of fresh water. Dispersion and diffusion of the salt-water component, as well as the density effect, were taken into account. A feature of this study was that different combinations of parameter values were tried in dimensionless form, resulting in four named parameters: seepage factor; dispersion-to-advection ratio; geometry ratio; and time-scale factor. The limitation of the sharp-interface approach in coastal aquifers for conditions of both steady state and unsteady state was also investigated in this study.

In order to mitigate seawater intrusion problems, Sherif and Hamza (2001) presented a two-dimensional finite element model (2D-FED) with a variable density flow to verify a technique for restoration of the balance between freshwater and saline water in coastal aquifers. Simulations were undertaken in the vertical view, and equiconcentration and equipotential lines were plotted for different locations of brackish water pumping. Results indicated that brackish water pumping exerted a significant influence on the width of the dispersion zone and that the quality of the pumped water was related to the pumping location. The method was applied to the Madras aquifer in India.

Choi *et al.* (2002) delineated a 2-D finite element model for the laterally unbounded density current developing on a slope, which solved the layer-averaged equations numerically using the Beam and Warming scheme. The computed flow profiles were compared with the numerical solution whilst the front velocity was compared with the measured data. The model was also applied to a laterally unbounded density current developing on a tilted surface.

Young *et al.* (2005) developed a two-dimensional laterally averaged reservoir model to study the density currents generated in a thermally stratified reservoir due to inflow with sediment concentration. The governing equations were solved using the Galerkin's weighted residual finite element method, with an arbitrary Lagrangian-Eulerian scheme. The model was also applied to simulate the field-scale hydrodynamic and temperature structures in the Te-Chi reservoir at Taiwan. Prediction of the different regimes of density currents such as the turbidity current, underflow and interflow were preformed for sediment-laden inflow during the days of heavy rainfall due to storms.

5.5 Three-dimensional models

Kodama *et al.* (2001) presented a verification of the newly improved multiple-level finite element model for application of 3-D tidal current analysis in Tokyo Bay. The improvement included additional effects due to various forcing factors, as well as a new numerical treatment of the open boundary condition in order to effectively eliminate the spurious reflective waves often generated by various numerical methods simulating free surface flows. Numerical experiments were conducted to carefully examine the tidal circulations affected by the interaction of the forcing factors, namely, Coriolis force, river inflows and wind shears. This study has also resulted in the enhancement of the accuracy of numerical simulations of three-dimensional flow in coastal waters by employing this model.

Chen *et al.* (2003) developed an unstructured grid, three-dimensional primitive equation ocean model for the study of coastal oceanic and estuarine circulation, which was applied to the Bohai Sea and the Satilla River. The irregular bottom slope was represented using a sigma-coordinate transformation, and the horizontal grids comprised unstructured triangular cells. The model combines the advantages of a finite element method for geometric flexibility and a finite difference method for simple discrete computation. It was concluded that currents, temperature, and salinity in the model computed in the integral form of the equations could provide a better representation of the conservative laws for mass, momentum, and heat in the coastal region with complex geometry.

Escribano *et al.* (2004) presented a three-dimensional numerical model using the finite element method to diagnose coastal currents off Antofagasta, in a well-known upwelling region of northern Chile. Steady-state conditions were simulated under the absence of wind or with the dominant upwelling-favourable wind in the zone. A 3-D particle tracking program was further used to diagnose and provide deeper understanding of expected particle trajectories over a steady-state period.

Wai *et al.* (2004) described a 3-D finite element sediment transport model integrating waves and currents to continuously account for different-scale activities, with application to simulate actual situations in the Pearl River Estuary, China. The wave action equation, which took into account wave refraction and diffraction as well as tidal hydrodynamic modification, was employed to compute the wave parameters. During the time marching process, the computation of the wave and current forcing was coupled in order that the effects due to short-term activities could be taken into account.

Ghostine *et al.* (2008) studied the numerical resolution of the three-dimensional Saint Venant equations for the study of flood propagation. A discontinuous finite element space discretization with a second-order Runge–Kutta time discretization was used to solve the governing equations. It was found that the explicit time integration coupled with the use of orthogonal shape functions rendered the method computationally more efficient than comparable second-order finite volume methods.

Abualtayef *et al.* (2008) presented the development and application of a three-dimensional multilayer hydrostatic model of tidal motions in the Ariake Sea, which was able to compute wetting and drying in tidal flats. The governing equations were derived from 3-D Navier–Stokes equations and were solved using the fractional step method, integrating the finite difference method in the horizontal plane and the finite element method in the vertical plane.

5.6   Characteristic-Galerkin method

5.6.1   *Formulation of the discretized equations*

The hydrodynamic and mass transport equations, when written in the fully conservative form, have the form:

$$\frac{\partial U}{\partial t} + \frac{\partial F_i}{\partial x_i} = R \tag{5.1}$$

In the above U is the vector of independent variables, which depends on the position vector x and the time t; F_i is the advective flux vector in the *i*th space coordinate and is a function of U and x. R is a self-adjoint operator on U that contains zero (source) and second-order (diffusion) space derivatives and may also depend on x. The vector R can be written as:

$$R = R_s + \frac{\partial R_{di}}{\partial x_i} \tag{5.2}$$

where R_s is a source term and R_{di} denotes the diffusion fluxes.

A second-order Taylor series expansion is developed for U in time about $t = t^n$ in the form

$$U^{n+1} = U^n + \Delta t \left(\frac{\partial U}{\partial t} \right)^n + \frac{\Delta t^2}{2} \left(\frac{\partial^2 U}{\partial t^2} \right)^n \tag{5.3}$$

The time derivatives can be replaced in terms of space derivatives as follows:

$$\frac{\partial U}{\partial t} = R - \frac{\partial F_i}{\partial x_i} = R_s + \frac{\partial R_{di}}{\partial x_i} - \frac{\partial F_i}{\partial x_i} \tag{5.4}$$

and

$$\frac{\partial^2 U}{\partial t^2} = \frac{\partial}{\partial t}\left(R_s + \frac{\partial R_{di}}{\partial x_i} - \frac{\partial F_i}{\partial x_i}\right) = G\frac{\partial U}{\partial t} + \frac{\partial}{\partial t}\left(\frac{\partial R_{di}}{\partial x_i}\right) - \frac{\partial}{\partial x_i}\left(\frac{\partial F_i}{\partial t}\right) \quad (5.5)$$

where $G = \partial R_s / \partial U$.

Since only approximations possessing C^0 continuity are considered, all derivatives higher than second order have to be dropped. The time-discretized equation is then

$$U^{n+1} = U^n + \Delta t\left(R_s + \frac{\partial R_{di}}{\partial x_i} - \frac{\partial F_i}{\partial x_i}\right)^n$$
$$+ \frac{\Delta t^2}{2}\left\{G\left(R_s - \frac{\partial F_i}{\partial x_i}\right) - \frac{\partial}{\partial x_i}\left[A_i\left(R_s - \frac{\partial F_j}{\partial x_j}\right)\right]\right\}^n \quad (5.6)$$

with $A = \partial F / \partial U$.

To complete the discretization process we shall consider the following approximations:

$$U = U_i N_i, F_j = F_j^i N_i, R_s = R_s^i N_i$$

where N_i is the piecewise linear shape function associated with node i, and

$$R_{dj} = R_{dj}{}^e P_e, G = G^e P_e, A_j = A_j{}^e P_e$$

where P_e is the piecewise constant shape function associated with element e. Figure 5.1 shows the discretization of linear elements in a solution domain in finite element method and Figure 5.2 shows the linear shape function N_i corresponding to a triangular element ijk. Equation (5.6) is then weighted with the shape functions N_j to give

$$(M\Delta U)_i = \Delta t \int_\Omega \left(R_s + \frac{\partial R_{di}}{\partial x_j} - \frac{\partial F_j}{\partial x_j}\right)^n N_i d\Omega + \frac{\Delta t^2}{2}\int_\Omega \left[G\left(R_s - \frac{\partial F_j}{\partial x_j}\right)\right.$$
$$\left. - \frac{\partial}{\partial x_j}\left\{\left[A_j\left(R_s - \frac{\partial F_k}{\partial x_k}\right)\right]\right\}^n N_i d\Omega \right. \quad (5.7)$$

where Ω is the problem spatial domain; $\Delta U = U^{n+1} - U^n$; and the entries in the consistent mass matrix M are given by

$$M_{ij} = \int_\Omega N_i N_j d\Omega \quad (5.8)$$

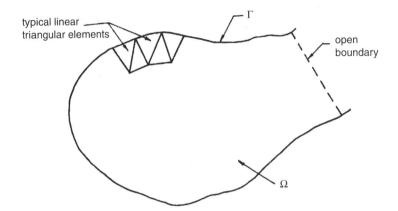

Figure 5.1 Discretization of linear elements in a solution domain in the finite element method

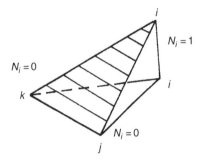

Figure 5.2 Linear shape function N_i corresponding to a triangular element *ijk*

Applying Green's theorem to equation (5.7), we have

$$
\begin{aligned}
(M\Delta U)_i = \Delta t & \left\{ \int_{\Omega} \left[\left(R_s - \frac{\partial F_j}{\partial x_j} \right)^n N_i + R_{di}^n \frac{\partial N_i}{\partial x_j} \right] d\Omega + \int_{\Gamma} R_{dj}^n n_j N_i d\Gamma \right\} \\
+ \frac{\Delta t^2}{2} & \left\{ \begin{array}{l} \int_{\Omega} \left\{ \left[G \left(R_s - \frac{\partial F_j}{\partial x_j} \right) \right]^n N_i + \left[A_j (R_s - \frac{\partial F_k}{\partial x_k}) \right]^n \frac{\partial N_i}{\partial x_j} \right\} d\Omega - \\ \int_{\Gamma} \left[A_j \left(R_s - \frac{\partial F_k}{\partial x_k} \right) \right]^n n_j N_i d\Gamma \end{array} \right\}
\end{aligned}
\quad (5.9)
$$

with n_j being the components of the outward unit normal to the boundary Γ.

The algorithm is in essence a finite element implementation of the Lax–Wendroff finite difference scheme used in high-speed flow computations

(Roache 1976). The main, and essential, difference lies, however, in the consistent mass matrix M, which links the contribution of neighbouring nodes; in a finite difference context the matrix M is "lumped", i.e. diagonal. For computational reasons it is nevertheless convenient to use a lumped form M_L, and the full effects M can be obtained by a simple iteration (Donea 1984). Thus the solution of the discretized equation (5.9) can be implemented as written

$$M\Delta U = f^n \tag{5.10}$$

and an iterative solution is obtained from

$$M_L \Delta U^k = f^n - (M - M_L)\Delta U^{k-1} \tag{5.11}$$

with $\Delta U^0 = 0$; three iterations are generally sufficient.

5.6.2 *Two-step algorithm*

The algorithm described above possesses excellent accuracy characteristics and it has been shown to work well in practice. However, when it is applied to systems of equations, it has the disadvantage of requiring the evaluation and subsequent multiplication of the matrices A and G. These operations are very time consuming and their vectorization will require the storing of these matrices for each element. In order to avoid this, the computation of the right hand side of equation (5.9) can be reorganized using the following two-step algorithm, which reduces the overall computational cost to approximately a half for the shallow water equations, and even less, when these are solved in a coupled manner with additional equations such as mass transport.

Considering the Taylor series expansion correct to first order and neglecting diffusion effects:

$$U^{n+1/2} = U^n + \frac{\Delta t}{2}\left(R_s - \frac{\partial F_i}{\partial x_i}\right) \tag{5.12}$$

Approximate $U^{n+1/2}$ in a piecewise constant manner with U^n, R_s and F_i interpolated as before. A suitable weighted residual form of equation (5.12) is then

$$\int_\Omega U^{n+1/2} P_e \, d\Omega = \int_\Omega U^n P_e \, d\Omega + \frac{\Delta t}{2}\int_\Omega \left(R_s - \frac{\partial F_i}{\partial x_i}\right)^n P_e \, d\Omega \tag{5.13}$$

This expression leads immediately to the distribution of U and completes the first step.

Again use Taylor series expansions to write

$$F_i^{n+1/2} = F_i^n + \frac{\Delta t}{2}\left(\frac{\partial F_i}{\partial t}\right)^n = F_i^n + \frac{\Delta t}{2}\left[A_i\left(R_s - \frac{\partial F_j}{\partial x_j}\right)\right]^n \tag{5.14}$$

$$R_s^{n+1/2} = R_s^n + \frac{\Delta t}{2}\left(\frac{\partial R_s}{\partial t}\right)^n = R_s^n + \frac{\Delta t}{2}\left[G\left(R_s - \frac{\partial F_j}{\partial x_j}\right)\right]^n \tag{5.15}$$

Expressions (5.14) and (5.15) can now be used to construct an alternative representation for the undesirable terms appearing in the one-step scheme. The distributions of $F_i^{n+1/2}$ and $R_s^{n+1/2}$ are approximated in a piecewise constant fashion by using directly the form of $U^{n+1/2}$ calculated in the first step, whereas F_i^n and R_s^n are interpolated as before. Weighting equations (5.14) and (5.15) with piecewise constant shape functions P_e, the following relations are found

$$\left[A_i\left(R_s - \frac{\partial F_j}{\partial x_j}\right)\right]^e = \frac{2}{\Delta t}\left(F_i^{n+1/2} - \overline{F}_i^n\right) \tag{5.16}$$

and

$$\left[G\left(R_s - \frac{\partial F_j}{\partial x_i}\right)\right]^e = \frac{2}{\Delta t}\left(R_s^{n+1/2} - \overline{R}_s^n\right) \tag{5.17}$$

where the overbar denotes the average value over the element e.

Using these expressions, equation (5.9) can be rewritten as

$$(M\Delta U)_i = \Delta t\left\{\int_\Omega\left[\left(R_s - \frac{\partial F_j}{\partial x_j}\right)^n N_i - R_{di}^n\frac{\partial N_i}{\partial x_j}\right]d\Omega + \int_\Gamma R_{dj}^n n_j N_i d\Gamma\right.$$

$$\left.+ \int_\Omega\left[\left(R_s^{n+1/2} - \overline{R}_s^n\right)N_i + \left(F_j^{n+1/2} - \overline{F}_j^n\right)\frac{\partial N_i}{\partial x_j}\right]d\Omega - \int_\Gamma\left(F_j^{n+1/2} - \overline{F}_j^n\right)n_j N_i d\Gamma\right\}$$

$$\tag{5.18}$$

5.6.3 A characteristics-based approach

The time-discretized equation (5.6) can also be obtained by following a characteristics approach. This will be shown for the 1-D scalar advection equation. Consider the transport equation, now written as

$$\frac{\partial\phi}{\partial t} + A\frac{\partial\phi}{\partial x} = 0 \tag{5.19}$$

with A being a function of ϕ only, i.e. $A = A(\phi)$.

Along the characteristic lines,

$$\frac{\mathrm{d}x}{\mathrm{d}t} = A \qquad (5.20)$$

equation (5.19) reduces to

$$\frac{D\phi}{Dt} = 0 \qquad (5.21)$$

or $\phi = $ constant. Thus, starting from a given initial distribution and a given spatial discretization, equation (5.21) may be used to construct the solution at later times with a continuous mesh updating via equation (5.19).

This computational drawback of mesh updating can be avoided as follows. Introduce a characteristic coordinate ξ, such that ξ is a constant along a characteristic, and consider the situation over a time interval $t^n \leq t \leq t^{n+1}$. Figure 5.3 shows the characteristic $\xi = $ constant which passes through the point P with coordinates (x_p, t^n) and the point Q with coordinates (x, t^{n+1}). We can write

$$\varphi(x, t^{n+1}) = \varphi(x_p, t^n) \qquad (5.22)$$

and, using (5.20),

$$x_p = x - \Delta t A_p \qquad (5.23)$$

with A_p being the value of A at P.

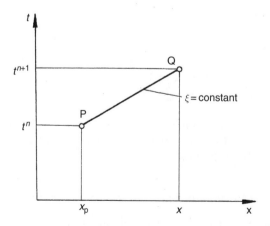

Figure 5.3 Path of a characteristic over the time interval (t^n, t^{n+1})

Using (5.23) in (5.22) and expanding as a Taylor series results in

$$\varphi^{n+1} = \varphi^n + \Delta t A_p \left(\frac{\partial \varphi}{\partial t} \right)^n + \frac{\Delta t^2}{2} A_p^2 \left(\frac{\partial^2 \varphi}{\partial t^2} \right)^n \qquad (5.24)$$

with

$$\varphi^{n+1} = \varphi \left(x, t^{n+1} \right) \qquad (5.25)$$

and

$$\varphi^n = \varphi \left(x, t^n \right) \qquad (5.26)$$

correct to second order. The value of A_p is approximated also by performing a Taylor series expansion

$$A_p = A - \Delta t A \left(\frac{\partial A}{\partial x} \right)^n \qquad (5.27)$$

with

$$A = A \left(\varphi^n \right) \qquad (5.28)$$

Combining (5.27) and (5.24) we obtain

$$\varphi^{n+1} = \varphi^n - \Delta t A \left(\frac{\partial \varphi}{\partial x} \right)^n + \frac{\Delta t^2}{2} \left\{ A^2 \left(\frac{\partial^2 \varphi}{\partial t^2} \right)^n + 2A \left(\frac{\partial A}{\partial x} \right)^n \left(\frac{\partial \varphi}{\partial x} \right)^n \right\} \qquad (5.29)$$

or, rearranging,

$$\varphi^{n+1} = \varphi^n - \Delta t A \left(\frac{\partial \varphi}{\partial x} \right)^n + \frac{\Delta t^2}{2} \frac{\partial}{\partial x} \left(A^2 \frac{\partial \varphi}{\partial x} \right) \qquad (5.30)$$

It can be observed that this is the same result that would be produced by applying equations (5.6)–(5.19). Thus the characteristics nature of the methodology presented has been shown.

5.6.4 *The conservative hydrodynamic and mass transport equations*

In order to exploit the full power of the numerical technique presented in the preceding section, it is essential to write the hydrodynamic and mass transport equations in conservative form. Then, in the absence of diffusive mechanisms, the equations can be written as:

$$\frac{\partial U}{\partial t} + \frac{\partial F_1}{\partial x_1} + \frac{\partial F_2}{\partial x_2} = R_s \tag{5.31}$$

which, integrated over a given spatial domain Ω, yields after application of Gauss's divergence theorem

$$\frac{\partial}{\partial t}\left(\int_{\Omega} U d\Omega\right) = -\int_{\Gamma}\left(F_1 n_1 + F_2 n_2\right)d\Gamma + \int_{\Omega} R_s d\Omega \tag{5.32}$$

This equation states that for any arbitrary domain, the ratio by which the total amount of U increases is given by the integral of the normal flux F_n over the boundary Γ of the domain Ω, plus the amount of U generated inside the domain (i.e. R_s). The components of the vector U are the physical conserved quantities: mass, momentum and any additional quantity which is being solved simultaneously with the shallow water dynamics (e.g. pollutant, concentration, temperature, etc.).

The usual forms of the hydrodynamic and mass transport equations found in the literature (Dronkers 1964; Benque *et al.* 1982) may not satisfy such requirements even in those cases where the equation is called conservative. Even if they are conservative, they may not be accurate. The equation presented here is equivalent in the differential form to the standard long-wave shallow water equation. The essential difference, however, lies in the different grouping of the various terms.

The tidal hydrodynamic and mass transport equations in the fully conservative form read:

$$\frac{\partial U}{\partial t} + \frac{\partial F_1}{\partial x} + \frac{\partial F_2}{\partial y} = R_s + \frac{\partial R_{d1}}{\partial x} + \frac{\partial F_{d2}}{\partial y} \tag{5.33}$$

with

$$U = \begin{bmatrix} h \\ hu \\ hv \\ hc \end{bmatrix}, F_1 = \begin{bmatrix} hu \\ hu^2 + 1/2g(h^2 - H^2) \\ huv \\ huc \end{bmatrix},$$

$$F_2 = \begin{bmatrix} hv \\ huv \\ hv^2 + 1/2g(h^2 - H^2) \\ hvc \end{bmatrix}$$

$$R_s = \begin{bmatrix} 0 \\ g(h-H)\partial H/\partial x - gu/C_z^2(u^2+v^2)^{1/2}+fhv \\ g(h-H)\partial H/\partial y - gv/C_z^2(u^2+v^2)^{1/2}-fhu \\ 0 \end{bmatrix} \tag{5.34}$$

$$R_{d1} = \begin{bmatrix} 0 \\ 2\mu h/\rho(\partial u/\partial x) \\ \mu h/\rho(\partial v/\partial x + \partial u/\partial y) \\ hD(\partial c/\partial x) \end{bmatrix}, R_{d2} = \begin{bmatrix} 0 \\ \mu h/\rho(\partial v/\partial x + \partial u/\partial y) \\ 2\mu h/\rho(\partial v/\partial y) \\ hD(\partial c/\partial y) \end{bmatrix}$$

and then can be solved by a characteristic Galerkin scheme, which has two distinct advantages: i) the small storage requirements render it suitable for microcomputers; and ii) it offers the vectorization properties and allows the coupled hydrodynamics and transport equations to be solved simultaneously.

5.6.5   *Accuracy analysis of advection-dominated problems*

We consider the general one-dimensional advection-dominated equation:

$$c_t + uc_x = 0 \tag{5.35}$$

where c may be regarded as the concentration of a material substance. Assuming constant velocity and coefficients, the accuracy of a time-stepping numerical scheme can be analysed using a procedure similar to that adopted in long-wave computations (Leendertse 1967).

The general solution to equation (5.35) is written as a Fourier series of the following form:

$$c = \sum_{n=-\infty}^{\infty} C_n e^{i(\beta_n t + \sigma_n x)} \tag{5.36}$$

where β_n is the frequency of the nth Fourier component, $\sigma_n = 2\pi/L_n$ is the wavenumber, L_n is the wavelength, and $i = \sqrt{-1}$. On substitution of equation (5.36) into (5.35) the following advection relationship between β_n and σ_n can be obtained:

$$\beta_n = -u\sigma_n \tag{5.37}$$

Because equation (5.35) is linear, we need only consider one Fourier component. From (5.36) and (5.37) we obtain the ratio of the analytical solution at $t+\Delta t$ to that at time t (eigenvalue) as:

$$\lambda = \frac{c(t+\Delta t)}{c(t)} = |\lambda| e^{i\theta} = e^{-iu\sigma\Delta t} \tag{5.38}$$

where the amplitude decay is 1 in this case and the exponential describes the translation of the analytical wave.

The error committed by the numerical solution of an advection-dominated problem described by (5.35) can be calculated. Adopting the CG method, the governing equation is first discretized in time:

$$c^{n+1} = c^n - u^n \Delta t c_x^n + \frac{1}{2} (u^n \Delta t)^2 c_{xx}^n \tag{5.39}$$

Using linear elements for spatial discretization, application of the Galerkin method gives the following discretized equations for an interior node i:

$$\frac{1}{6} \left(c_{i+1}^{n+1} + 4c_i^{n+1} + c_{i-1}^{n+1} \right) = \frac{1}{6} \left(c_{i+1}^n + 4c_i^n + c_{i-1}^n \right) - \frac{1}{2} Cr \left(c_{i+1}^n - c_{i-1}^n \right)$$
$$+ \frac{1}{2} Cr^2 \left(c_{i+1}^n - 2c_i^n + c_{i-1}^n \right) \tag{5.40}$$

where $c_i^n = c(i\Delta x, n\Delta t)$, and $Cr = u\Delta t/\Delta x$ is the Courant number.

Noting that $c_{i+1}^n = c_i^n e^{i\sigma \Delta x}$, $c_{i-1}^n = c_i^n e^{-i\sigma \Delta x}$, the numerical eigenvalue, $\lambda^* = c_i^{n+1}/c_i^n = |\lambda^*|e^{i\theta^*}$ can be obtained as:

$$\lambda^* = 1 + \frac{3Cr^2 (\cos p - 1)}{(\cos p + 2)} - i \frac{3Cr \sin p}{(\cos p + 2)} \tag{5.41}$$

where $p = \sigma \Delta x = 2\pi \Delta x/L$.

The scheme is stable if $|\lambda^*| \le 1$, i.e.

$$\left[1 + \frac{3Cr^2 (\cos p - 1)}{(\cos p + 2)} \right]^2 + \left[\frac{3Cr \sin p}{(\cos p + 2)} \right]^2 \le 1 \tag{5.42}$$

or, after simplification,

$$Cr \le 1/\sqrt{3} = 0.57 \tag{5.43}$$

This gives a stability criterion which is more restrictive than that of conventional explicit finite difference schemes, $Cr \le 1$.

The numerical error per time step is measured by the complex ratio $P = \lambda^*/\lambda$, which depends on the Courant number of the scheme and the wavelength of interest. The amplitude ratio $|P|$ gives the degree of damping or amplification, while the relative celerity θ^*/θ measures the phase error relative to the analytical wave.

The numerical propagation factors P and θ^*/θ for the characteristic-Galerkin scheme are plotted in Figure 5.4 and Figure 5.5 respectively as a function of the number of grid points per wavelength $L/\Delta x$ for representative Courant numbers of 0.1, 0.3 and 0.5. The amplitude ratio converges

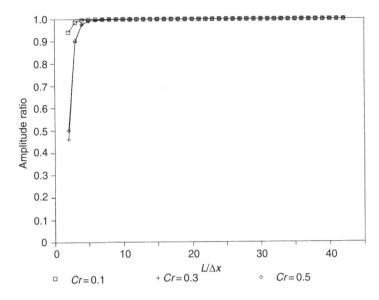

Figure 5.4 Amplitude of the characteristic-Galerkin method for the pure advection
equation for different Courant numbers

to unity for all wavelength as the Courant number approaches zero. From
Figure 5.4 it can be seen that for $Cr \leq 0.3$, the scheme displays minimal
amplitude error for $L/\Delta x \geq 5$. The characteristic-Galerkin scheme is slightly
damped for the short waves. Considerable insight can be gained by exam-
ining the phase characteristics in Figure 5.5. It is interesting to note that
the characteristic-Galerkin method has optimal phase characteristics at a
Courant number of $Cr \approx 0.3$; a slight phase lead of the short waves is
noted. It has been proven in Lee *et al.* (1987) that the overall performance
of the numerical propagation characteristics of the characteristic-Galerkin
method is superior to those of the related classical Lax–Wendroff method
and the implicit Crank–Nicolson scheme and is comparable to that of a
characteristic-based finite difference scheme which uses Hermitian cubic
interpolating polynomials (Holly and Preissmann 1977). The numerical
calculations performed as part of this study were carried out based on a
modified version of a code originated at the Institute for Numerical Method
in Engineering, University College of Swansea (Peraire *et al.* 1986).

5.7 Verification of the numerical scheme

Before the model can be reliably applied to prototype problems, the stability
and accuracy of the numerical schemes are studied in extensive analytical
tests with idealized geometries. Model results are compared against exact

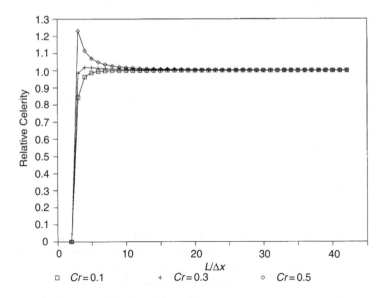

Figure 5.5 Phase portraits of the characteristic-Galerkin scheme for the pure advection equation for different Courant numbers

analytical solutions. Under these circumstances guidelines on the appropriate choice of mesh size and time step can be developed. In this section, we present detailed test results in many applications. Special attention is placed on computed velocities, which are more difficult to simulate than wave heights. The accuracy of the velocity results depends, amongst other factors, on how faithfully the bottom topography is represented.

The scheme is first subjected to five relevant two-dimensional analytical tests which embrace many essential realistic features of environmental and coastal hydrodynamic applications: pure advection of a Gaussian hill, pure rotation of a Gaussian hill, advective diffusion in a plane shear flow, continuous source in tidal flow, and long waves entering a rectangular channel with quadratic bottom bathymetry. The numerical results are compared with exact solutions. For the first three test cases, the numerical results are also compared with those of a fractional step method (FS method) which has been proven with unconditional L_∞ stability and with insignificant numerical damping effect in a steady-state test case (Wu and Chen 1985; Wu 1986). The FS method employs linear elements, a lumped mass matrix solution and linear interpolation of characteristics.

5.7.1 Pure advection of a Gaussian hill

The scheme is asked to purely advect a Gaussian concentration distribution at a constant velocity $u = 0.5$ m/s in a uniform long channel, i.e. $D = 0$ m^2/s.

The sharp Gaussian distribution is described by only ten computational points on a uniform grid with $\Delta x = \Delta y = 200\,\text{m}$.

The initial concentration is given by

$$c_0(x) = \exp\left(-\frac{(x-2000)^2}{2\sigma_0^2}\right) \qquad (5.44)$$

where $\sigma_0 = 264\,\text{m}$.

The exact solution is

$$c(x,t) = \exp\left(-\frac{(x-2000-ut)^2}{2\sigma_0^2}\right) \qquad (5.45)$$

At a Courant number of 0.24, the numerical solutions by the CG and FS method are taken at $t = 9600\,\text{s}$, corresponding to a transported distance of 4.8 km. For the CG scheme, the peak concentration is reduced by only about 10 per cent at $t = 9600\,\text{s}$. The results reveal minor node-to-node oscillations in the y-direction. Both the results at $y = \Delta y$ and at $y = 2\Delta y$ perform similarly. For the FS method, the corresponding peak concentration is reduced by as much as 70 per cent. It is demonstrated that the FS method gives unacceptably large numerical damping when the Peclet number $(u\Delta x/D)$ is large.

5.7.2   *Pure rotation of a Gaussian hill*

This problem concerns the transport by convection of a 2-D Gaussian concentration-hill in a flow in anticlockwise rigid body rotation. The mathematical problem is governed by the equation

$$\frac{\partial c}{\partial t} + u\frac{\partial c}{\partial x} + v\frac{\partial c}{\partial y} = 0 \qquad (5.46)$$

with initial and boundary conditions

$$c(x,y,0) = c_0(x,y)$$
$$c(x,y,t) \to 0 \quad \text{as } x^2 + y^2 \to \infty$$

where

$c(x,y,t)$ is the concentration field,
$u(y) = -\omega y$ is the x-velocity,
$v(x) = \omega x$ is the y-velocity,
ω is the angular frequency of rotation $(=2\pi/3000)$, period $= 3000\,\text{s}$,

$c_0(x, y)$ is the initial concentration field, defined as

$$c_0(x, y) = \exp\left(-\frac{x^2}{2\sigma_0^2} - \frac{(y - 1800)^2}{2\sigma_0^2}\right) \tag{5.47}$$

$(0, 1800)$ is the centre of mass of the initial concentration field, and σ_0 is the standard deviation of the initial concentration field ($= 264$ m). The exact solution is of the form

$$c(x, y, t) = \exp\left(-\frac{(x - ut)^2}{2\sigma_0^2} - \frac{(y - 1800 - vt)^2}{2\sigma_0^2}\right) \tag{5.48}$$

A uniform grid size ($\Delta x = \Delta y = 200$ m) is employed. The numerical solution is performed in a 2-D grid defined as follows:

$$x, y \in [-3400, 3400]$$
$$x(i, j) = 200(i - 1) - 3400 \qquad i = 1, 35$$
$$y(i, j) = 200(j - 1) - 3400 \qquad j = 1, 35$$

where $x(i, j)$, $y(i, j)$ are the nodal coordinates (indices i and j refer to the x- and y-directions respectively). The time step Δt chosen is 10 s. The advection number ($u_{max}\Delta t/\Delta x$) based on the maximum velocity is around 0.3. All of these and the following calculations were performed on a microcomputer.

Comparison of the computed results was made after a quarter revolution for a rotation period of 3000 s. Figure 5.6 shows the three-dimensional view of the computed result. For the CG scheme, although the concentration is described by only about ten grid points, the peak concentration is dissipated by only 10 per cent, with minimal phase distortion at the tails. For the FS scheme to deal with advection-dominated problems, the numerical damping is too excessive (about 80 per cent). This may be related to the linear back-tracking of characteristics and mass lumping of this particular scheme (Wu 1986). This is evident in the unrealistically large eddy viscosity coefficient reported in their tidal circulation simulation of the Bohai Sea.

5.7.3 Advective diffusion in a plane shear flow

This problem concerns the transport of small sources in a plane shear flow with diffusion. The problem was presented by Carter and Okubo (1965) and related discussion can be found in Okubo and Karweit (1969).

For the simple case of a steady 2-D unidirectional flow, with velocity u_0 along the x-axis, shear $\lambda = du/dy$, and a constant diffusion coefficient D, the governing equation is

$$\frac{\partial c}{\partial t} + (u_0 + \lambda y)\frac{\partial c}{\partial x} = D\left[\frac{\partial^2 c}{\partial x^2} + \frac{\partial^2 c}{\partial y^2}\right] \tag{5.49}$$

Concentration

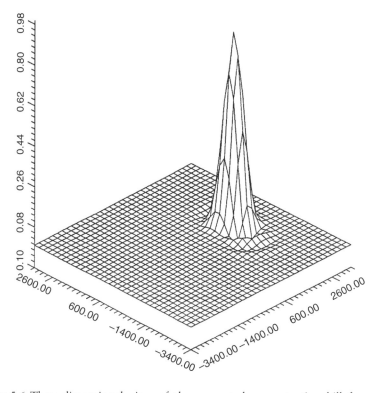

Figure 5.6 Three-dimensional view of the computed concentration hill for pure advection of a rotating cone

subject to initial and boundary conditions of

$$c(x, y, 0) = c_0(x, y)$$

$$c(x, y, t) \to 0 \qquad \text{as } |x| \text{ or } |y| \to \infty$$

When the initial condition is a point source of mass m at $x = x_0 = 7200$, $y = y_0 = 0$, and $t = 0$, the solution is

$$c(x, y, t) = \frac{m}{4\pi Dt(1 + \lambda^2 t^2/12)^{1/2}} \exp\left(-\frac{(x - x' - 0.5\lambda yt)^2}{4Dt(1 + \lambda^2 t^2/12)} - \frac{y^2}{4Dt}\right)$$

$$(5.50)$$

where $x' = x_0 + u_0 t$.

To allow numerical solution based on a finite source size, calculation should begin at time $t = t_0$ with a concentration distribution given by the above equation using $m = 4\pi D t_0 (1 + \lambda^2 t_0^2/12)^{1/2}$ i.e. the initial peak concentration should be unity. In this example, t_0 is taken to be 2400 s.

Computation is made for the advective diffusion of an initial finite source in a plane shear flow, $u(y) = u_0 + \lambda y$ at time $t = 4800$ s, where $u_0 = 0.5$ m/s, $\lambda = 5.0 \times 10^{-4}$, and the diffusivity $D = 10$ m^2/s. The grid size is 200 m ($\Delta x = \Delta y = 200$ m) and the time step is 96 s. The advection number based on the maximum velocity is around 0.3. The computed and exact solutions are in excellent agreement. It is again shown that the accuracy of the CG scheme is far better than that of the FS scheme.

5.7.4 Continuous source in a tidal flow

The governing equation for this problem is as follows:

$$\frac{\partial c}{\partial t} + (u_f + u_t \sin 2\pi t/T)\frac{\partial c}{\partial x} = D\frac{\partial^2 c}{\partial x^2} - kc + S \tag{5.51}$$

The concentration distribution resulting from a step steady injection of mass at $x = 0$, $t = 0$, in a time-varying flow is considered (Li and Lee 1985). The advective x-velocity consists of a net downstream component in the $+x$ direction u_f, and a reversing sinusoidal component of amplitude u_t and period T:

$$u(t) = u_f + u_t \sin(2\pi t/T) \tag{5.52}$$

The steady forcing term $S(x,t) = S(x)$ is modelled by the linear shape function corresponding to the source node at $x = 0$. The analytical periodic steady-state solution corresponding to a finite source for a representative case, using $u_f = 0.03$ m/s, $u_t = 0.61$ m/s, $T = 12.4$ hr, $D = 6$ m^2/s, $k = 0.034$/day is performed. The computed solution at both high water slack (HWS) and low water slack (LHS) by the CG scheme is shown in Figure 5.7 for a representative case, using $\Delta t = T/24$ and $\Delta x = 3050$ m. Starting from $c(x,0) = 0$, the computed steady state reached in about 90 cycles is essentially described by an equilibrium concentration profile which oscillates up and down the channel with the phase of the advective flow. The solution at $t = 90T$ is compared with the exact solution corresponding to the triangular-shaped source forcing. It is seen that the results with the CG method, without any grid refinement near the source, are in very good agreement with the analytical solution. Further, the results show negligible oscillations around the upstream front of the concentration distribution.

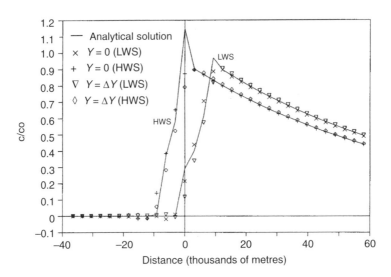

Figure 5.7 Comparison of concentration profile by CG method and analytical solution at high water slack (HWS) and low water slack (LWS)

5.7.5 Long wave in a rectangular channel with quadratic bottom bathymetry

The linearized shallow water equations with bottom frictional dissipation can be written as:

$$\eta_t + (uh)_x + (vh)_y \,, u_t + g\eta_x + \tau u = 0, v_t + g\eta_y + \tau v = 0 \tag{5.53}$$

where u, v are the velocities in the x- and y-directions, η is the free surface elevation above the undisturbed level, and τ is the linear friction coefficient.

The analytical solution to long wave propagation in a rectangular channel with quadratic bottom bathymetry has been given by Lynch and Gray (1978):

$$z(x,t) = \mathrm{Re}\left\{ \left[Ax^\alpha + Bx^\beta\right] e^{\frac{i2\pi t}{T}} \right\} \tag{5.54}$$

$$u(x,t) = \mathrm{Re}\left\{ \left[A\alpha x^{(\alpha-1)} + B\beta x^{(\beta-1)}\right] \frac{i2\pi}{\psi^2 H_0 T} e^{\frac{i2\pi t}{T}} \right\} \tag{5.55}$$

where $i =$ unit complex vector

$$A = a\beta x_1^\beta / (\beta x_1^\beta x_2^\alpha - \alpha x_1^\alpha x_2^\beta)$$

$$B = -a\alpha x_1^\alpha / (\beta x_1^\beta x_2^\alpha - \alpha x_1^\alpha x_2^\beta)$$

$A = $ amplitudeofsurfaceoscillationattheopenend

$$\alpha, \beta = -0.5 \pm (0.25 - \psi^2)^{1/2}$$

$$h(x) = H_0 x^2$$

$$\psi^2 = 2\pi \frac{2\pi/T - \tau i}{TgH_0}$$

A standing oscillation in uniform channel of length $l = 10$ km with quadratic bottom bathymetry of mean water depth varying from 12 m at the open end to 3 m at the closed end is considered. Figure 5.8 shows the elevation of the schematized channels. A surface sinusoidal oscillation of amplitude $a = 0.5$ m and period $T = 44,640$ s is prescribed at the open end. At a solid boundary, the normal velocity is made to vanish, except at the corner nodes where the no-slip condition is imposed. A constant linearized bottom friction coefficient of $\tau = 0.100 \times 10^{-3}$ /s is assumed. This example is solved

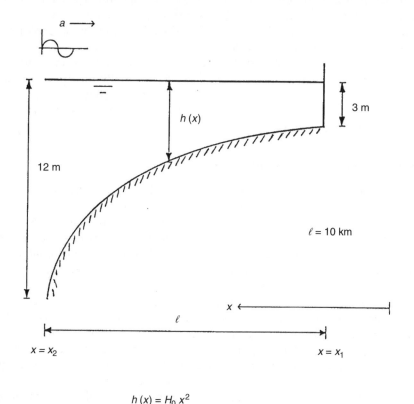

$$h(x) = H_0 x^2$$

Figure 5.8 Elevation view of schematized channel with quadratic bottom bathymetry and mean water depth ranging from 3 m to 12 m

by the two-dimensional version of the CG method written in conservative form.

By employing the CG method, the channel is divided into ten segments of uniform dimension ($\Delta x = \Delta y = 1000$ m). The finite element grid comprises 40 triangular elements and 33 nodes. The wavelength can be estimated as $L = T\sqrt{(gh)} = 484 \sim 242$km $> 4^{*}$l and $L/\Delta x = 242 \sim 484$. A "cold start" condition is employed as the initial value to the problem. Starting from zero elevations and velocities, the periodic steady-state is reached in about 5–6 cycles. At a Courant number of 0.4, the computed periodic steady-state water elevation and water velocities in the 10 km quadratic bottom bathymetry test case are compared with the analytical solution. The agreement with the exact solution is very good.

Extensive comparisons of the results have been performed with analytical test problems which embrace many essential realistic features of environmental and coastal hydrodynamic applications: pure advection of a Gaussian hill, pure advection of a rotating cone, advective diffusion in a plane shear flow, dispersion of a continuous source in a tidal flow, and long-wave propagation with bottom frictional dissipation in a channel with quadratic bottom bathymetry. They serve as a check on the developed codes and give an insight into the choice of Courant number in prototype applications of similar dimensions.

However, it is oversimplified to consider real estuaries in the above forms which can be solved analytically. Real estuaries usually incorporate non-linear effects of complicated topography and bed frictional resistance so that it is worthwhile investigating the effectiveness of the scheme in the cases of real prototype applications.

5.8   Advantages and disadvantages

When compared to the finite difference method, the finite element method can be more easily adapted to handle cases with irregular geometries.

As the algorithms presented are explicit and solve steady-state or periodic problems by a semi-dynamic relaxation procedure, the storage requirements are small, and the method can be used on mini- and microcomputers if necessary. This aspect is of some importance to practising engineers who may wish to make use of such computations.

Because linear elements are used, exact integration can be easily carried out. Therefore numerical integration is not needed. This represents a considerable saving in computer time. Moreover, the restriction in the Courant criterion limit is more than compensated for by the gain in accuracy. The scheme is particularly attractive in situations where the Courant stability constraint does not result in an excessively small time step.

The algorithms are fully vectorizable, thus allowing the user to take advantage of the dramatic speed-up factors which can be obtained on the new generation of supercomputers. The equations are written in vector form, which means that additional effects such as pollutant dispersion or temperature distribution can be included in a straightforward manner. If desired, additional equations modelling the hydrodynamic transport of any number of scalar variables can be solved together with the shallow water equations. The additional scalar transport equations can indeed be considered in the general formulation by only adding additional components to the unknown, flux and source vectors.

The hydrodynamic and mass transport equations are written in a fully conservative form. The complete derivation starting from the Navier–Stokes equations is presented. The main difference with respect to the existing hydrodynamic and mass transport equations lies in the grouping of the pressure terms, which in the standard formulations are represented in the primitive form. This fully conservative form, together with the numerical scheme employed, leads to an algebraic system of equations which preserves, in a discrete manner, the conservation of the mass and momentum. The solution algorithm is based on an explicit time integration procedure which exploits the conservative properties of the governing equations and incorporates accurate treatment of convective terms with minimal numerical damping. Checks have been made on the mass conservation of the scheme for both one-dimensional and two-dimensional test cases. For the test case of pure advection of a Gaussian hill, the ratio of total masses to initial masses in the whole domain remains almost unity after 100 time steps.

5.9 Prototype application I: mariculture management

In previous sections, it has been demonstrated that the numerical model is capable of solving the hydrodynamic and mass transport equations correctly. In this section, the characteristic-Galerkin method is used to study a prototype problem: tidal flushing for two fish culture zones located in Tolo Harbour, Hong Kong. Simulations are first performed with the mean tide used as the forcing function at the open boundary; the dynamic steady-state flow field and tidal flushing rates at the two fish culture zones are computed. The coupled hydrodynamic and mass transport equations are solved for seven days, about half a spring-neap cycle. The open boundary condition is provided by a synthetic tide of 42 tidal constituents derived from an extended harmonic analysis of long-term records. The predictive ability of the model is assessed.

5.9.1 *General description of Tolo Harbour*

Tolo Harbour, lying in the north-eastern part of the New Territories, Hong Kong, is a nearly land-locked water body with a tidal inlet to Mirs Bay in the eastern waters. It has an area of about $52 \, \text{km}^2$ and extends about $20 \, \text{km}$ from south-west to north-east with an entrance width of less than $1.5 \, \text{km}$.

The elongated inlet, Tolo Channel, connects the harbour to Mirs Bay, which is one of the major south-facing bays along the South China coast.

The water depth at the entrance to Tolo Harbour is about 25 m, but tapers off to about 2 m along the south-west and north-west shoreline where the new towns of Shatin and Taipo are located. Over 1,000 hectares of the harbour have been or are being reclaimed to form part of the two new towns, which will have a predicted population of about 850,000 on completion. Therefore, a rather heavy pollution loading will be exerted upon this region. Due to its enclosed nature, the waste assimilation capacity is very limited. Of special interest is the computation of flushing rates for the two fish culture zones located in the inner coves of Yim Tin Tsai West and Three Fathoms Cove (Figure 5.9).

Despite the slight stratification observed during the wet season, Tolo Harbour can be considered a vertically fully-mixed estuary for most of the year, and a two-dimensional vertically averaged model can be applied. The ratio of average freshwater inflow per tidal period to tidal prism is in the order of 0.01 and the freshwater inflow has an insignificant effect upon the dynamics of the typical harbour circulation. Therefore, the flow is mainly driven by the tidal forcing at the entrance to the harbour.

From the tidal records, it can be observed that tides in this area are predominantly semi-diurnal of the mixed type. The tides are often distorted due to amplification of the shallow water constituents. Therefore the observed tide is more complicated than a simple harmonic tide, and is composed of

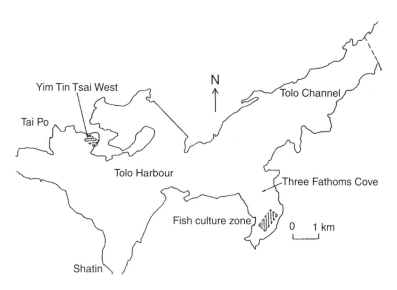

Figure 5.9 General layout of Tolo Harbour

a number of harmonic tidal constituents with period down to about three hours.

5.9.2 Dynamic steady-state simulation: M2 tidal forcing

The first step in the numerical simulation is to construct the spatial grid for the region concerned. Laying out a good grid is the most fundamental aspect of finite element modelling and is essential for an efficient solution. This process requires skill on behalf of the modeller, which can be acquired only through experience. The following guidelines are adopted (Wang and Connor 1975; Norton and King 1976):

(a) The problem's overall boundaries are sketched out first with as few corners as possible. It is wise to begin any new problem with the crudest approximation possible and then later make refinements as they become necessary. Since some inaccuracies are to be expected in the data, the boundary should be reasonably far away from any area of interest.
(b) Once the network's overall limits have been set it is usually advisable to construct a series of lines more or less parallel to the long axis of the problem. The position and frequency of these lines should reflect the areas of special importance in the problem, the expected hydraulic gradients, and the degree of relief in the bottom contours.
(c) Once the longitudinal lines have been located, transverse lines can be drawn, again recognizing the requirements and details of the specific problem. After the basic grid has been established, the triangular structure can be filled in with special attention to corners and other areas where gradients are expected to be large.
(d) Grid dimensions should change gradually, and for accuracy the elements must not degenerate. For triangular elements, no apex angle should approach zero and preferably they should be almost equilateral.

A compromise must be found between accuracy and computational efficiency. The grid size is chosen under such requirements that it is small enough to provide adequate resolution for the physical phenomena concerned, but large enough to keep the computational effort required to the minimum. In this case, a grid size of about 1,000 m is generally used in Tolo Harbour and the grid formed contains 93 unknown nodal points and 125 linear triangular elements. A finer grid of 500 m is adopted in the two bays of interest: Three Fathoms Cove and Yim Tin Tsai West. Five grid points are also located consistent with the fish culture zone covering the four corners and the middle point with grid size down to 200 m (Figure 5.10).

Generally, hydrodynamics and mass transport are coupled problems, and the scalar transport equation can indeed be considered in the general formulation by adding only one additional component to the unknown, flux and source vectors. In practice, however, the time scales of the two phenomena

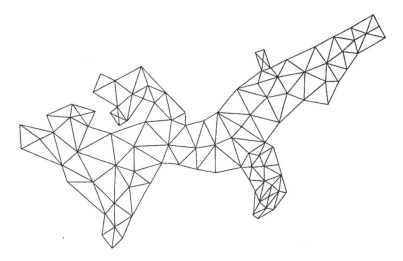

Figure 5.10 Spatial grid of Tolo Harbour

can be very different, in which case the coupling becomes very weak. Based on the stability consideration, the allowable time step for the transport equation is 465 s, 93 times that of the shallow water equation. This represents an unnecessary expense if the whole set of equations (hydrodynamics and transport) is to be solved simultaneously.

The method adopted here is to solve the whole set of equations in a semi-coupled form. As the allowable time step for the transport equation is 93 times that for the current calculation, the tidal current computation is advanced 93 steps before the advective diffusion equation is solved; by the nature of the explicit scheme, this can easily be achieved. At this stage the solution has been advanced by a time step of 465s, and the transfer of information between hydrodynamics and transport can be performed before repeating the same process again. In this manner, the transport equation is not solved every time step, and the essential effects of the coupling are still retained.

As the mean freshwater inflow is very small, they are neglected in the simulations. Thus only two types of boundary conditions are required. At the land boundary, the forcing is given by the known time history of the elevation. For the grid chosen, the only forcing is at the ocean boundary located at the entrance to Tolo Channel where the width is small, and the tilting of the water surface along the open boundary can be assumed to be small and safely neglected.

For the dynamic steady-state simulation, a semi-diurnal sinusoidal M2 forcing with an amplitude equal to 0.85 m and a period equal to 44,640.0 s (12.4 hr) is used to represent the mean tide in Tolo Harbour, which has an average tidal range of 1.7 m. Based on the stability considerations, a time step of 5 s resulting in a Courant number of 0.25 is used in all steady-state simulations. As there is no prior knowledge about the flow field, all the simulations are "cold-started", i.e. all the variables are assumed to be zero initially. The adopted values of the Manning's coefficient and the eddy viscosity coefficient are 0.035 and 50 m²/s respectively.

With the selected model coefficients, the tidal circulation in Tolo Harbour caused by a pure sinusoidal tide is studied. The results clearly indicate a near-standing wave character for the tidal propagation, and velocities are approximately 90° out of phase with the surface elevation variation. The phase lag between the ocean boundary and the interior at Taipo and Shatin is only about 5–8 min. Also the amplification of the wave amplitude is very small. The computed tidal currents are in general agreement with field observations. It was found that after one cycle the results tend to repeat, and maximum currents of about 0.3 m/s and 1–3 cm/s are found in the outer harbour and inside the weakly flushed fish culture zone respectively.

Using the computed tidal current of a repeating tide as input, the advective diffusion is also solved. Lateral turbulent mixing and vertical shear dispersion are modelled by a gradient-type relation, and the lumped dispersion coefficient is evaluated locally as $D = |u|h$, where $|u|$ is the absolute value of the computed time-varying velocity and h is the local mean depth. In fact, the results obtained by assuming $D = 0.6|u|h$ are similar; the results are not sensitive to the exact value of the dispersion coefficient.

A unit mass of conservative tracer/pollutant is initially discharged from the middle of each fish culture zone, and the tracer mass inside the zone, affected by both tidal advection and dispersion, is tracked for ten tidal cycles. When the pollutant was released in the fish culture zone in Three Fathoms Cove, the tidal flushing effect and the mass in the adjacent segments at high water slack are computed. It is found that the exponential rule is followed after about seven tidal cycles and the flushing rate is determined to be about 0.13/day with flushing time equalling eight days. The results are consistent with the field observation data. The flushing rate at Yim Tin Tsai West is much weaker.

5.9.3 Real tide simulation for seven days (42 tidal constituents)

The flushing rates of the two zones are compared by solving the coupled hydrodynamic and tracer mass transport equations simultaneously for seven days, about half a spring-neap cycle. The open boundary condition is provided by a synthetic tide of 42 tidal constituents derived from an extended harmonic analysis of long-term records.

With initial concentrations specified as discussed above, the concentration field is subsequently tracked for seven days. Pollutant decay is neglected in this test case. Mass conservation is maintained at the landward end; at the seaward end, no conditions are imposed during ebb, while a prescribed concentration, equal to a fraction of the average value during the previous ebb, is prescribed. A sensitivity analysis has been carried out to study the effect of this fraction on the results. It is found that the variation of this fraction from 0.3 to 0.5 has an insignificant effect on the result and hence a value of 0.5 is adopted.

The magnitude of the ebb current inside Three Fathoms Cove ranges from 0.01 m/s to 0.03 m/s during neap tide and spring tide, respectively. The Lagrangian pathlines in Three Fathoms Cove are also computed. It is observed that the fishermen, by their practical experience, also align their fish rafts within the fish culture zone roughly in this direction. The mass in the two fish culture zones is continuously tracked (Figure 5.11); an exponential decay can be discerned, and the flushing rate for the fish culture zone in Three Fathoms Cove is around 0.1 per day, consistent with observations. The flushing rate at Yim Tin Tsai West is much weaker.

By employing the robust CG model, a successful simulation has been made of the realistic tidal variation and scalar transport in Tolo Harbour. The flushing rates of semi-enclosed bays in the harbour have been determined via a solution of the full equations over half a spring-neap cycle. The numerical results can be gainfully applied in water quality modelling and mariculture management. All the computations are performed on a microcomputer. The run time for the prototype case, with 7,452 time steps per periodic tidal cycle, is 20 minutes for each cycle. Since the grid is rather coarse in most of Tolo Harbour, the above serves only as an initial demonstration of the method and a preliminary calculation. The computational

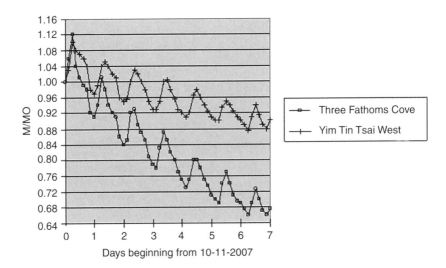

Figure 5.11 Tidal flushing computation: variation of tracer mass in two fish culture zones over a spring-neap cycle

results are to be ascertained by a full grid. More refined calculations with the use of an automatic mesh generator are needed. A finer grid, constructed by the application of an automatic mesh generation technique, is adopted in the next prototype application.

5.10   Prototype application II: the effect of reclamation on tidal current

In the last section, the mathematical model was applied to mariculture management in Tolo Harbour, i.e. a land-locked water body. In this section, the model is applied to study the effect of the proposed massive reclamation under the Hong Kong Port and Airport Development Strategy project (PADS) on tidal current in Victoria Harbour, Hong Kong, which is a difficult task since there are two open boundaries.

5.10.1   General description of Victoria Harbour

Victoria Harbour is a tidal bay located between Hong Kong Island and Kowloon Peninsula. Two sections of Victoria Harbour are chosen as the model limit, and tidal forcings at both ends are imposed. The harbour has two tidal inlets from the South China Sea – one at Lei Yu Mun on the eastern side and the other adjacent to Stonecutter Island on the western side. The width varies from approximately 500 m at Lei Yu Mun to 6,000 m at the western end of the harbour. Figure 5.12 shows the location of the study area. It extends about 12 km from east to west; the water depth varies from 6 m along the shoreline to 21 m at Lei Yu Mun and adjacent to the central portion of the harbour.

5.10.2   Hydrodynamic simulation for an M2 tidal forcing

For modelling flow and transport processes in natural flow systems, the use of the finite element method with irregular triangular meshes finds wide application. The optimal configuration of a finite element mesh for a particular flow problem depends to a greater or lesser degree upon each of the following three criteria: adequate resolution of bathymetry, hydrodynamical considerations, numerical accuracy. Since the manual generation of finite element meshes which simultaneously comply with each of the above criteria is a cumbersome and extremely difficult task, computer assistance in the automatic generation, refinement and smoothing of such meshes is highly desirable.

In recent years, much attention has been given to the automatic generation of irregular computational grids. A review of various methods is given by Thacker (1980). Although the application of some techniques has proved most successful as a means of generating meshes for constant depth systems, the case of irregular bathymetry introduces considerable problems owing

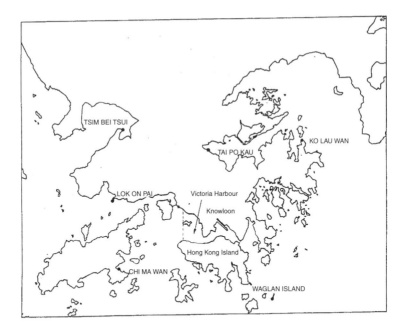

Figure 5.12 General layout of Victoria Harbour

to the often conflicting requirements of hydrodynamic and topographical criteria (Westwood and Holz 1986).

An automatic mesh generator (Lo 1991; Lo 1992; Lo and Lee 1994) is employed. Figure 5.13 shows the grid layout for Victoria Harbour. The grid formed contains 273 unknown nodal points and 455 linear triangular elements. A grid size of around 300 m is used for the main part of the harbour. The bathymetry of the harbour is determined and interpolated using linear shape functions. The bathymetry representation is acceptable, when compared with the original bathymetry of the harbour. Based on published tidal harmonic constituents, a mean M2 semi-diurnal tidal forcing of range 1.7 m with a tidal phase difference of 10° is imposed at the two open boundaries.

As there is no prior knowledge about the flow field, all the simulations are "cold-started", i.e. all the variables are assumed to be zero initially. The time step chosen is 5 s. In the model, the adopted values of Manning's n and the eddy viscosity coefficient are 0.025 and 50 m^2/s respectively.

Figure 5.14 shows an example of the computed maximum ebb current at a Courant number of approximately 0.3. A maximum current of around 0.5 m/s is found in the harbour at Lei Yu Mun. Similar magnitude of maximum currents has been found in the harbour by employing an alternating-direction-implicit finite difference model (Choi, Lee and Cheung 1989).

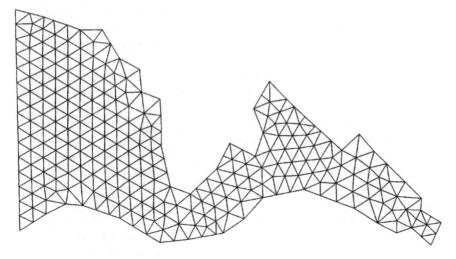

Figure 5.13 Spatial grid of Victoria Harbour by the automatic mesh generator

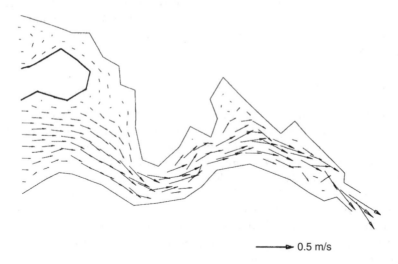

──────▶ 0.5 m/s

Figure 5.14 Computed maximum ebb current in Victoria Harbour for a mean tide ($\Delta\eta = 1.7$ m)

Comparison is also made with the Hong Kong Tidal Stream Atlas (Hydrographic Department 1975). It is found that the maximum tidal current in the Tidal Stream Atlas is, in general, larger. However, as no allowance is made for the diurnal inequality in the tidal streams, errors as great as ±1 knot may occur. A sensitivity analysis has been carried out to study the effect of different tidal phase differences imposed at the two open boundaries and the effect of different tidal amplitudes imposed at the two boundaries. It is found that both these factors will affect significantly the magnitudes of the computed velocities. Their effects are that larger tidal phase differences between the two open boundaries and a larger tidal amplitude at the Lei Yu Mun side contribute to larger tidal currents inside the harbour.

However, since it is found from the Tide Table (Hong Kong Observatory 1994) that the tidal phase differences between Lei Yu Mun and Tsuen Wan will not exceed 10°, this value is adopted. If the tidal amplitude at Lei Yu Mun is higher than that at the western end by 10 per cent, the agreement between the computed velocities and the tidal stream atlas is much better. Numerical experiments have been performed and comparisons of tidal velocities over a periodic cycle have been made between the computed results and the tidal stream atlas at a few control points inside the harbour, i.e. Lei Yu Mun, Kowloon Bay, Tsimshatsui and Stonecutter Island.

It is noted that the magnitude of the tidal current is greatest at Lei Yu Mun, becomes smaller near Kowloon Bay and adjacent to Tsimshatsui, and then becomes weak near Stonecutter Island at the western end of the harbour. It is also observed that there is an almost constant phase difference between the computed tidal velocities and those published in the tidal stream atlas at different locations. The phase comparison has been reconciled by a shift in time origin. It is concluded that both the magnitude and phase of the computed tidal current agree well with those published in the tidal stream atlas, to an accuracy permitted by the latter.

5.10.3 *Real tide simulation for four principal tidal constituents*

From the published tidal constituents, it can be observed that tides in this area are predominantly semi-diurnal of the mixed type (F number equals 1.17). Tidal forcing with the four principal tidal harmonic constituents at Quarry Point and at Tsuen Wan, as presented in Table 5.1, is imposed at the two open boundaries. It is also noted that mean sea level is different at the two open boundaries. The mean sea level at Quarry Point is 1.38 m above chart datum whilst that at Tsuen Wan is 1.47 m above chart datum. In order to give a realistic tidal pattern of the harbour, the numerical model is run for three days around the time of spring tide and for another three days around the neap tide in January 1987.

The maximum tidal currents during a typical spring tide and neap tide are 0.6 m/s and 0.4 m/s respectively at Lei Yu Mun, whilst the maximum velocity during a mean M2 tidal forcing from the above section is 0.5 m/s at the same place. The results obtained seem to be reasonable.

5.10.4 *Effect of reclamation*

The total population of Hong Kong is just under six million, with about 60 per cent concentrated around the Victoria Harbour area. Because the

Table 5.1 Tidal harmonic constants at Quarry Bay and Tsuen Wan

Location	Mean tide level (mCD)	Harmonic constituent	Local amplitude (m)	Local phase (rad)	Frequency (rad/hr)
Quarry Bay	1.38	O_1	0.34	1.4266	0.2434
		K_1	0.40	1.2254	0.2625
		M_2	0.39	1.2494	0.5059
		S_2	0.16	1.0647	0.5236
Tsuen Wan	1.47	O_1	0.33	1.2870	0.2434
		K_1	0.41	1.2254	0.2625
		M_2	0.42	1.0748	0.5059
		S_2	0.18	0.9948	0.5236

upland areas are difficult to develop and are prone to landslides, much of the densely populated existing urban areas is on low-lying land created through coastal land reclamations in the past 100 years. Since 1841, drastic changes of the coastline have taken place within Victoria Harbour. New reclamations are currently being constructed and are being planned for the Port and Airport Development Strategy (PADS).

Figure 5.15 shows a map of Victoria Harbour showing the position of the coastline during 1990 and the coastline proposed for PADS. Since the area of the proposed reclamation is massive in comparison with the total water area in Victoria Harbour (~1/3), it is of interest to study its effect on tidal flows inside the harbour.

From the previous section, it has been shown that the tidal velocity can be well represented by a mean M2 tidal forcing. The harbour after reclamation is subjected to the same mean M2 tidal forcing at the two open boundaries. An assumption is made that the proposed reclamation will not affect the tidal level or the phase imposed at the two open boundaries.

After the proposed reclamation, the new grid formed contains 201 unknown nodal points and 316 linear triangular elements. A grid size of around 300 m, similar to the previous case, is used for the main part of the harbour. A mean M2 semi-diurnal tidal forcing of range 1.7 m with a tidal phase difference of 10° is imposed at the two open boundaries. All the simulations are "cold-started". The same parameters (time step, Manning's *n*, eddy viscosity coefficient, etc.) are adopted as in the previous case. A maximum current of order of magnitude 0.2 m/s is found in the harbour at Lei Yu Mun. Figure 5.16 shows the comparison of tidal velocities over a periodic cycle before and after the proposed reclamation at a control point inside the harbour, i.e. Lei Yu Mun.

It is observed that the tidal current is reduced significantly by about 50 per cent at Lei Yu Mun and near Kowloon Bay due to the proposed massive reclamation. The reduction in magnitude becomes less (about 30 per cent) adjacent to Tsimshatsui whilst the effect is only very slight near Stonecutter

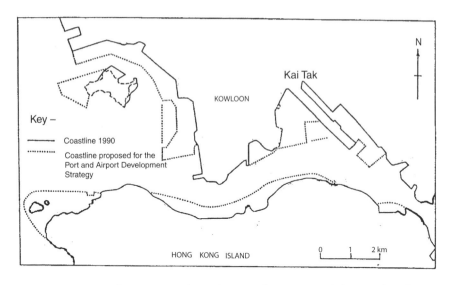

Figure 5.15 Map of Victoria Harbour showing the position of the coastline during 1990 and the coastline proposed for PADS

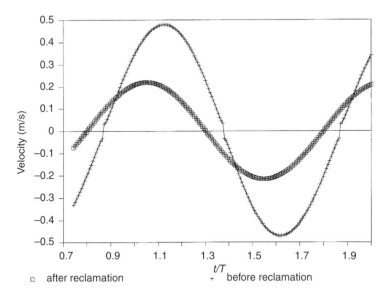

Figure 5.16 Effect of reclamation on tidal velocities over a periodic cycle at Lei Yu Mun

Island (about 10 per cent). It is also seen that the proposed reclamation will affect the phase of the tidal current at Lei Yu Mun (about one hour) whilst its effect is very slight at the other control locations. This significant decrease in velocity is alarming, and remains to be further examined using a larger computational domain. A case with deeper water depths around Lei Yu Mun (30 m) has been tested and it is found that the computed tidal current is not related to the water depths.

5.11 Conclusions

It is desirable to be able to predict, accurately and efficiently, water levels and currents as well as pollutant transport in tidal estuaries under realistic synthetic tidal boundary condition and over at least half a spring-neap cycle. Appropriate coastal development and environmental measures can be formulated based on these computations. In coastal waters changes in water quality often occur over much larger time scales than those of tidal current. Hence, an effective way is needed to interface between hydrodynamic and water quality models.

The governing equations of hydrodynamic and mass transport are utilized to describe unsteady constant density flow in a water body which may be subject to tidal forcing, upstream freshwater inflows and/or pollutant release.

An explicit characteristic-based Galerkin finite element scheme suited for advection-dominated problems has been analysed and implemented for the solution of the coupled hydrodynamic and mass transport equations. It is shown for the 1-D scalar advection equation that the time-discretized equation can also be obtained by following a characteristics approach. An understanding of the computational stability, accuracy and efficiency has been gained from an analysis of the numerical scheme. As the scheme uses linear elements, exact integration can be carried out. The algorithm is fully vectorizable: thus the power of the new generation of supercomputer can be fully exploited. Also, this renders the coupling of tidal current and mass transport straightforward. The small storage requirement also renders the scheme suitable for microcomputer calculations.

The accuracy of the numerical solution is tested against several 2-D representative analytical solutions which embrace many essential realistic features of environmental and coastal hydrodynamic applications: pure advection of a Gaussian hill, pure rotation of a Gaussian hill, advective diffusion in a plane shear flow, continuous source in a tidal flow and long waves entering a rectangular channel with quadratic bottom bathymetry. Excellent agreement has been recorded with the analytical solution.

The effectiveness of the numerical modelling of estuaries can only be demonstrated by practical applications as well as by validation with field data. The model has been applied to study hydrodynamics and mass transport in two harbours with different characteristics: Tolo Harbour, which is semi-enclosed, and Victoria Harbour, which has two open boundaries. It is proved that the characteristic-Galerkin scheme is able to reproduce the observed characteristic physical behaviour of tidal dynamics as well as pollutant transport.

The results demonstrate the efficiency and accuracy of this robust explicit finite element scheme for the solution of 2-D hydrodynamic and mass transport problems. All the computations have been performed on a micro-computer which is readily available in a design office. There is much potential for the practical us of the model in solving prototype problems.

Future developments of this topic include: a) a more detailed study of tidal flushing time and effect of reclamation on tidal hydraulics; b) an investigation of numerical procedures to partially remove the Courant stability restriction; and to treat tidal flats; and c) the use of the model to study water quality in 2-D laterally averaged density-stratified estuarine flows.

6 Soft computing techniques

6.1 Introduction

During the past few decades, physically based or process models based on mathematical descriptions of water motion such as those described in the last few chapters have been widely used in coastal management. Conventionally, the emphasis on computer-aided decision-making tools has been placed primarily on their algorithmic processes, and in particular on the formulation of new models, improved solution techniques, and effectiveness (Leendertse 1967). In a water quality model, which addresses a typical coastal problem, phytoplankton dynamics are based on theories of the dependence of growth and decay factors on physical and biotic environmental variables (e.g. solar radiation, nutrients, flushing) – expressed mathematically and incorporated in advective diffusion equations. Classical process-based modelling approaches can give a good simulation of the water quality variables including algal biomass level, but usually require a lengthy data calibration process. They require a lot of input data and rely upon many uncertain kinetic coefficients. They sometimes make simplified approximations of various interrelated physical, chemical, biochemical, and biological processes (Sacau-Cuadrado et al. 2003; Patel et al. 2004; Arhonditsis and Brett 2005). Difficulties are often encountered in modelling coastal waters with limited data on the water quality and the cost of water quality monitoring. While a large amount of research has already taken place in numerical modelling, a complementary approach to corroborate the numerical results should be welcome owing to computational limitations and uncertainties in modelling the complex physical process of coastal dynamics.

Moreover, the current technique for the numerical simulation of water resources has become a highly specialized task, involving certain assumptions and/or limitations, and can be manipulated only by experienced engineers who have a thorough understanding of the underlying theories. This has led to significant constraints on the use of models, thus generating a discrepancy between the developers and users of models. The models are often not user-friendly enough. They lack the ability to transfer knowledge in the application and interpretation of the model, to offer expert support to novice users, and to achieve effective communication from developers to users. Many users of a model do not possess the requisite knowledge to glean their input data, build algorithmic models, and evaluate the results of their model. The result may be the production of inferior designs and the under-utilization or even total failure of these models. Recently, there has been an increase in demand for an integrated approach. Thus, the problem is to present the information, knowledge, and experience in a format that facilitates comprehension by a broad range of users from novices to experts (Abbott 1989).

During the past decade, the information revolution has fundamentally altered the traditional planning, modelling, and decision-making methodologies of water-related technologies. The recent advances in artificial intelligence (AI) technologies are making it possible to integrate machine learning (ML) capabilities into numerical modelling systems in order to bridge the gaps and lessen the demands on human experts. Information technology now plays an essential role in the sustainable development and management of water resources. In addition, the general availability of sophisticated personal computers with ever-expanding capabilities has given rise to increasing complexity in terms of computational ability in the storage, retrieval, and manipulation of information flow. In this context, hydroinformatics, which only received its proper name in 1991 (Abbott 1991), is a new and emerging technology that has now become one of the most important branches of research and application in hydraulics and water resources (Chen *et al.* 2006). Hydroinformatics has been defined broadly to be the application of modern information technologies to the solution of problems pertinent to the aquatic environment, comprising integration of traditional fields of computational hydraulics with novel developments in information technology and computer science. It comprises the integration of information acquired from diverse sources, from field data to data from hydraulic and numerical models, to data from non-engineering fields such as economics, ecology, and even social science (Odgaard 2001). The First International Conference on Hydroinformatics was held in 1994 (Verwey *et al.* 1994), and successive conferences in the same series have since been organized every two years (Müller 1996; Babovic and Larsen 1998; Iowa Institute of Hydraulic Research 2000; Falconer *et al.* 2002; Liong *et al.* 2004; Goubesville *et al.* 2006; Universidad de Concepción 2009). In parallel, the *Journal of Hydroinformatics*, focusing on contemporary developments in this hot topic, has been published since 1999.

In the last decade soft computing techniques have become more and more popular. These novel models rely upon the methods of computational intelligence and machine learning, and thus assume the presence of a considerable amount of data delineating the underlying physics or phenomenon of the modelled system. Expert knowledge is frequently incorporated in mechanistic models to link environmental conditions to the hydrodynamic and water quality parameters. Since the formalization of problem-specific human expert knowledge is often difficult and tedious, data-driven machine learning techniques are a feasible alternative to extract knowledge from field datasets. It can be shown that data-driven models can complement knowledge-based expert approaches and hence improve model reliability. A general introduction to some contemporary soft computing techniques is provided in this chapter.

6.2 Soft computing

Soft computing is one of the latest approaches for the development of systems that possess computational intelligence. Soft computing techniques, including artificial neural networks, fuzzy logic, knowledge-based systems (KBS) and evolutionary algorithms, employ different computing paradigms to handle dynamic, non-linear and noisy data. They are especially useful when the underlying physical relationships are not fully understood (Zadeh 1994). Soft computing is a powerful data-modelling tool that is able to capture and represent highly non-linear complex input/output relationships and to complement physics-based models. Soft computing techniques are algorithms that estimate hitherto unknown mapping (or dependence) between a system's inputs and its outputs from the available data. Once a dependence is discovered, it can be used to predict (or effectively deduce) the system's future outputs from the known input values. The use of soft computing is particularly an option when, for example, the physical world is not fully defined, the model certains many uncertainties (model coefficients, boundary conditions, input parameters etc), it is extremely difficult and time-consuming to develop an accurate analytical model based on known mathematical and scientific principles, and/or there is a high cost involved in large-scale water quality monitoring.

Soft computing is computationally very fast and requires far fewer input parameters and input conditions than deterministic models. These novel techniques are ideally suited to model coastal dynamics since such models can be set up rapidly and are known to be effective in handling dynamic, non-linear and noisy data, especially when underlying physical relationships are not fully understood, or when the required input data needed to drive the process-based models are not available (Chen *et al.* 2006). Soft computing techniques are particularly useful when some information and/or data are missing. When compared with the full set of data required by process-based models, soft computing requires only a good number of representative data for training purpose. Numerical experiments can be done readily to determine the key variables and the optimal number of dependent variables. A potential application of a trained data-driven model is to provide its simulated values at desired locations where no measured data are available and yet are required for water quality models. This model could be used as a new prediction tool, which complements the process-based model, to identify the important parameters for enabling selective monitoring of water quality parameters and quick water quality assessment.

The new technologies with the widest applicability in the field are artificial neural networks (ANN), support vector machines (SVM), particle swarm optimization (PSO), data mining, knowledge-based systems, fuzzy systems, genetic algorithms (GA) and genetic programming (GP). Applications of these innovative techniques have recently been recorded in the literature. Chen *et al.* (2008) presented a comprehensive review of the use of artificial intelligence techniques for modelling environmental systems. Lin *et al.* (2008) employed an integrated hydrodynamic and ANN modelling approach to predict faecal indicator levels in estuarine receiving waters. Sylaios *et al.* (2008) developed a spreadsheet tool for the fuzzy modelling of chlorophyll concentrations in coastal lagoons in Greece with a raw dataset consisting of four predictor variables, i.e. water temperature, dissolved oxygen content, dissolved inorganic nitrogen concentration, and solar radiation levels. Preis and Ostfeld (2008) used a coupled model tree-genetic algorithm scheme for flow and water quality predictions in watersheds. Chen (2003) evaluated the trophic state of reservoirs by applying genetic programming. However, these applications were adopted only for a specific situation and might not be able to accomplish the desired performance under other circumstances. Moreover, different hybrid systems have started to emerge recently that can incorporate a variety of computer-aided tools to facilitate decision-making by model users. Talib *et al.* (2008) forecast algal dynamics in two shallow lakes by recurrent artificial neural networks and hybrid evolutionary algorithms. Zou *et al.* (2007) delineated an adaptive neural network embedded genetic algorithm approach for inverse water quality modelling. C.-F. Chen *et al.* (2008) determined optimal water resource management through a fuzzy multiobjective programming and genetic algorithm for a case study in Taiwan. Pinthong *et al.* (2009) developed a hybrid genetic and neurofuzzy computing algorithm to enhance the efficiency of water management for a multipurpose reservoir system in Thailand.

Chau (2006a) reviewed the state-of-the-art in the integration of different AI technologies into coastal modelling. The algorithms and methods reviewed comprise KBSs, genetic algorithms, artificial neural networks and fuzzy inference systems. Although KBSs had apparent advantages over other methods in rendering more transparent transfers of knowledge in the use of models and in providing the intelligent manipulation of calibration parameters, it is certain that other AI methods make their individual contributions towards accurate and reliable predictions of coastal processes. The integrated model might be very powerful, since the advantages of each technique can be coupled.

An ANN, constituted by highly interconnected neurons, is a massively parallel-distributed information processing system that mimics the human brain and nervous system. It uses processing elements connected by links of variable weights to form black box representations of systems. A typical

ANN comprises several layers of interconnected neurons, each of which is connected to other neurons in the ensuing layer. Data are presented to the neural network via an input layer, while an output layer holds the response of the network to the input. One or more hidden layers may exist between the input layer and the output layer. All hidden and output neurons process their inputs by multiplying each input by its weight, summing the product, and then processing the sum using a non-linear transfer function to generate a result. The data-driven models have the ability to learn complex model functions from examples. The major advantage is the ANN's capability to learn from the data but not require deep understanding of the process of the problem being studied. However, its drawback is its low transparency, resulting from the inability to interpret its internal working in a physically meaningful way.

Fuzzy logic is very useful in modelling complex and imprecise systems. Under fuzzy set theory, the elements of a fuzzy set are mapped to a universe of membership values using a function-theoretic form belonging to the close interval from zero to one. An important step in applying fuzzy methods is assessment of the membership function of a variable, which parallels the estimation of probability in stochastic models. Membership functions in fuzzy set theory, which are appropriate for modelling the preferences of the decision-maker, can be obtained on the basis of actual statistical surveys. Modelling based on fuzzy logic is a simple approach, which operates on an "if–then" principle, where "if" is a vector of fuzzy explanatory variables or premises in the form of fuzzy sets with membership functions, and "then" is a consequence also in the form of a fuzzy set. Fuzzy logic has been used in a number of applications but generally as a refinement to conventional optimization techniques in which the usual crisp objective and some or all of the constraints are replaced by fuzzy constraints. Fuzzy set theory concepts can be useful in water quality modelling, as they can provide an alternative approach to deal with those problems in which the objectives and constraints are not well defined or information about them is not precise. They can represent knowledge in a way that can be easily interpreted by humans. However, the number of rules will increase exponentially with an increase in the number of inputs and the number of fuzzy subsets per input variable.

Evolutionary algorithms (EA), such as GA and GP, are search techniques based on the mechanism of natural genetics and biologically inspired operations. GAs can be employed as an optimization method to minimize or maximize an objective function. GP differs from the traditional GA in that it typically operates on parse trees instead of bit strings. They apply the concept of the artificial survival of the fittest coupled with a structured exchange of information using randomized genetic operators taken from nature to compose an efficient search mechanism. This form of search evolves throughout iterative generations by improving the features of potential solutions and mimicking the natural population of biological creatures. EA can be applied to the evolution of models with more transparent knowledge representations, which facilitates understanding of model predictions and model behaviour. They may also help in determining the patterns, regularities and relationships which exist and drive a certain phenomenon, such as algal abundance. Through a variety of operations to generate an enhanced population of strings from an old population, they exploit useful information subsumed in a population of solutions. Various genetic operators that have been identified and used include crossover, deletion, dominance, intra-chromosomal duplication, inversion, migration, mutation, selection, segregation, sharing, and translocation. They hold the ability of global searching, yet may not necessarily lead to the best possible local solution.

A KBS is an interactive computer-based decision-making tool that emulates the intensive expert knowledge in a specific domain area. Typically it mimics the reasoning processes of experts, offers expert advice, and addresses domain problems requiring specific training. KBSs, which improve productivity and efficiency, can reduce the gap between a large workload and insufficient manpower. Basic components of a KBS are system context, knowledge base and inference engine. Additional components which are required to contribute a more functional system are knowledge acquisition, a user interface and explanation facilities. A knowledge-based decision support system, serving both as a design aid and as a training tool, will allow water engineers to become acquainted with up-to-date simulation tools and fill the existing gaps between researchers and practitioners in the application of recent technology in solving real prototype problems. This may lead to a better understanding of the advantages, applicability, and limitations of the different methodologies.

It can be observed that each of the above techniques has its own advantages and disadvantages, and that their performances depend highly upon each individual case. Recently, hybrid approaches (e.g. Loia *et al.* 2000; Zou *et al.* 2007; S.H. Chen *et al.* 2008; Talib *et al.* 2008) have also become possible. For example, Pinthong *et al.* (2009) employed a hybrid learning algorithm that combined the gradient descent and the least-square methods to train a genetic-based neurofuzzy network by adjusting the parameters of the neurofuzzy system. Although an individual algorithm has been mostly attempted in previous studies, it might also be feasible to develop hybrid systems incorporating a variety of tools that suit best in the particular situation.

It is not possible to start from scratch to develop advanced soft computing models. Some commercially available computer software and tools might be acquired in order to utilize the advances already made in this direction and to facilitate the development of the models. With rapid developments in recent years, many successful shells are now available. The usual requirements are the latest microcomputer-based soft computing shells that incorporate user-friendly interfaces, graphics capabilities and debugging tools under an Internet client/server network. It should have modular capabilities to interface with external programs in a Windows environment. Various kinds of Windows-type displays, such as form, checkbox group, list box, command button, textbox, option button, picture box, etc., should be available. The platform for developing the models is mainly the personal microcomputer, because of its widespread popularity in design offices. The prototype models thus developed will have the widest application in real-life situations.

6.3 Data-driven machine learning (ML) algorithms

During the past two decades, researchers have had at their disposal many fourth generation coastal models, ranging from numerical, mathematical and statistical methods to techniques based on AI. ML is an area of computer science, a sub-area of AI concentrating on the theoretical foundations (Solomatine 2002). An ML technique is an algorithm that estimates hitherto unknown mapping (or dependence) between a system's inputs and its outputs from the available data (Mitchell 1997). Once a dependence is discovered, it can be used to predict (or effectively deduce) the system's future outputs from the known input values. The growing development of computer-aided analysis, which is easily accessible to all researchers, has facilitated the application of various ML techniques in coastal modelling. These techniques include ANN (Recknagel *et al.* 1997; Chau and Cheng 2002), fuzzy and neuro-fuzzy techniques (Maier *et al.* 2001; Chen and Mynett 2003), evolutionary based techniques (Bobbin and Recknagel 2001; Jeong *et al.* 2003), etc. Although most of these studies are applied to freshwater environments (i.e., limnological or riverine systems), a few have been applied to saltwater eutrophic areas (Scardi and Harding 1999; Scardi, 2001; Lee *et al.* 2003; Muttil *et al.* 2004).

Recknagel *et al.* (1997) developed an ANN model for prediction of algal blooms in four different freshwater systems and compared its performance against several benchmarking conventional approaches. The water-specific limnological time-series comprised biomass of the ten dominating algae species as observed for 12 years and the measured environmental driving variables.

Yabunaka *et al.* (1997) delineated the application of a back propagation ANN model to predict algal bloom by forecasting growth of five phytoplankton species and the chlorophyll-a concentration in Lake Kasumigaura, Japan. Results illustrated that the ANN model was able to learn the relationship between the selected water quality parameters and algal bloom under appropriate training and validation.

Maier *et al.* (1998) employed backpropagation ANNs to make a four-week forecast of cyanobacteria *Anabaena* spp. in the River Murray at Morgan, Australia, on the basis of seven years of weekly data for eight variables. Satisfactory results were obtained in terms of both the incidence and magnitude of a growth peak of the cyanobacteria. Evaluations were also made on the use of lagged versus unlagged inputs, the predominant variables in determining the onset and duration of cyanobacterial growth.

Scardi and Harding (1999) delineated the development of a backpropagation ANN model for the prediction of primary production of phytoplankton in Chesapeake Bay, United States. Results comparisons of two multilayer perception architectures with different numbers of input variables were made with benchmarking conventional empirical models. Whilst the architecture with more numbers of input variables was demonstrated to be able to furnish more accurate results in this case study, the results of a sensitivity analysis were also presented.

Jeong *et al.* (2001) proposed a recurrent ANN for time-series modelling of phytoplankton dynamics and for predicting the timing and magnitudes of chlorophyll-a in the hypertrophic Nakdong River, Korea, on the basis of a four-year database at a study site located upstream of the river mouth. Meteorological, hydrological and limnological parameters were selected as input variables whilst chlorophyll-a concentration was the output variable. A sensitivity analysis was also undertaken to determine relationships amongst seasons, specific input variables and chlorophyll-a as well as the time lag of input data which would furnish the most accurate solution.

Scardi (2001) presented and discussed some approaches that could enhance neural network models to overcome the problem imposed by the limited amount of available data in phytoplankton primary production modelling in saltwater eutrophic areas. An adopted approach that could accomplish this purpose was to select additional inputs from a broader range of variables as copredictors. Moreover, information acquired from existing models could be effectively exploited and embedded into the ANN models by a metamodelling approach via a constrained training procedure.

Wei *et al.* (2001) developed an ANN model to predict the timing and magnitude of algal blooms in order to quantify the interactions between abiotic factors and algal genera in Lake Kasumigaura, Japan. Algal responses to the orthogonal combinations of certain external environmental factors, including chemical oxygen demand, pH, total nitrogen and total phosphorus, were simulated successfully. Specific combinations of environmental factors that would enhance the proliferation of some algae as well as other combinations that would inhibit bloom formation were found.

Maier *et al.* (2001) employed a B-spline associative memory networks (AMNs) fuzzy model to forecast concentrations of the cyanobacterium *Anabaena* spp. in the River Murray at Morgan, South Australia, four weeks in advance. This method enabled the information that was stored in trained networks to be expressed in the form of a fuzzy rule base. The performance of the model was evaluated and compared with that of a benchmarking ANN model, in terms of forecasting accuracy and model transparency. It was found that whilst the accuracy of the forecasts acquired employing the AMN was only marginally better, it had the advantage of furnishing more explicit information about the relationship between the model inputs and outputs.

Bobbin and Recknagel (2001) presented an application of an evolutionary algorithm to the problem of knowledge discovery on blue-green algae dynamics in a hypertrophic lake. Patterns in chemical and physical parameters of the lake and the corresponding presence or absence of highly abundant algae species were discovered by the machine learning algorithm. Learnt patterns were represented explicitly as classification rules, exhibiting hypothesized favourable environmental conditions for three different species of blue-green algae. Evaluation illustrated that models could be evolved which differentiate algae species based on the furnished environmental attributes.

Recknagel (2001) presented a preview of forthcoming developments in applications of machine learning to ecological modelling and projected that newly emerging adaptive agents were able to furnish a novel framework for the discovery and forecasting of emergent ecosystem structures and behaviours in response to environmental changes. He proposed that ANNs would be very useful for non-linear ordination and visualization of ecological data via Kohonen networks, and ecological time-series modelling via recurrent networks, whilst GAs would be innovative for hybridizing deductive models, and evolving predictive rules, process equations and parameters.

Recknagel *et al.* (2002) performed a more detailed comparison of potentials and accomplishments of ANNs and GAs in predicting algal blooms. They found that GAs outperformed ANNs for a case study in the eutrophic freshwater Lake Kasumigaura, Japan. Several advantages of employing GAs are noted including the ability to evolve, refine and hybridize numerical and linguistic models, higher accuracy in seven-days-ahead predictions of algal blooms, and provision of more transparent explanation.

The use of an accurate water stage prediction is to allow the pertinent authority to issue a forewarning of an impending flood and to implement early evacuation measures when needed. Existing methods including rainfall-runoff modelling or statistical techniques require exogenous input together with a number of assumptions. Chau and Cheng (2002) employed ANN to forecast real-time water levels in Shing Mun River of Hong Kong with different lead times according to the upstream gauging stations or stage/time history at the station itself. The network was trained by employing two different algorithms. It was shown that the ANN approach, which was able to furnish model-free estimates in deducing the output from the input, was a proper forewarning tool. It was seen from the training and verification modelling that the water elevation forecast results were highly accurate and were acquired with very small computational effort. Both these two factors were significant in water resources management. Also, sensitivity analysis was undertaken to evaluate the most appropriate network characteristics including the number of input neurons, number of hidden layers, number of neurons in a hidden layer, number of output neurons, learning rate, momentum factor, activation function, number of training epochs, termination criterion, etc. in this case study. The findings resulted in the reduction of any redundant data collection as well as the fulfilment of cost-effectiveness.

Różyński and Jansen (2002) employed a data-intensive principal oscillation pattern technique to set up a data-driven model of bed dynamics in order to analyse a non-tidal and mildly sloping nearshore zone at the Coastal Research Station Lubiatowo in Poland. Three reasonable patterns of long-term bed dynamics were generated. The modelling results were fairly accurate at bar locations where bed evolution was slow enough to be grasped by annual records. It was reported that, in terms of the explained variance, the POP model outperformed conventional empirical orthogonal functions modes.

Gournelos *et al.* (2002) studied the erosional process of north-eastern coastal Attica by the construction of an erosion risk map based on a web-geographic information system (GIS) and soft computing technology. The geology of this area was characterized by the alpine formation, with Mesozoic limestones and post-alpine deposits. Factors taken into consideration in the study included the effect of rapid urbanization during the last decade, the recent occurrence of a severe fire incident with enormous effect on the vegetation cover, the outcropping of post-alpine vulnerable formations which might accelerate erosion during possible intense rainfall, etc. The authors proposed that such an approach could become a helpful tool in regional planning and environmental management.

Lee *et al.* (2003) presented an ANN model with a back propagation learning algorithm to predict the algal bloom dynamics of the coastal waters of Hong Kong of a eutrophic sub-tropical nature. It was shown that results were quite in contrast to previous studies in freshwater systems by others, which suggested that more complicated neural networks of algal blooms would have better performance. They found that, in a eutrophic sub-tropical coastal water, the algal concentration was essentially dependent upon the antecedent algal concentrations during the previous one to two weeks, though it entailed a minimum sampling interval of one week.

Chen and Mynett (2003) developed a fuzzy logic (FL) model which coupled data mining techniques and heuristic knowledge to predict algal biomass concentration in the eutrophic Taihu Lake, China. The self-organizing feature map technique and empirical knowledge were applied jointly to define the membership functions and to induce inference rules. Results illustrated the potentials of exploring "embedded information" by coupling data mining techniques and heuristic knowledge.

Jeong *et al.* (2003) simulated the dynamics of bloom-forming algae in a eutrophic river–reservoir hybrid system at lower Nakdong River, Korea, using both a GP algorithm and multivariate linear regression (MLR). Results indicated that an inductive-empirical approach was more appropriate than MLR or a mechanistic model in mimicking the dynamics of bloom-forming algal species in a river–reservoir transitional system. The GP model was very successful in predicting the temporal dynamics and magnitude of blooms while MLR resulted in insufficient predictability.

Existing methods of water stage prediction including rainfall-runoff modelling or statistical techniques entail exogenous input together with a number of assumptions. The use of artificial neural networks has been found to be a cost-effective technique. But their training, usually with a backpropagation algorithm or other gradient algorithms, is found to have some drawbacks, including very slow convergence and easily getting stuck in a local minimum. Chau (2004b) developed a particle swarm optimization (PSO) model to train perceptrons, which was shown to be feasible and effective by forecasting real-time water levels in the Shing Mun River of Hong Kong with different lead times according to the upstream gauging stations or stage/time history at the station itself. It was shown from the verification simulations that faster and more accurate results can be obtained. Chau (2004c) presented a hybrid split-step PSO model for training perceptrons. It was shown that the results were both more accurate and faster to accomplish, when compared with the benchmark backward propagation algorithm and the original PSO algorithm.

Choudhury *et al.* (2004) employed an ANN-based model, with backpropagation learning, for classifying the occurrence and non-occurrence of seasonal thunderstorms over the eastern coastal region of India. Certain types of thunderstorms might possess great potential to cause serious damage to human life and property. In their study, soft computing techniques were used to forecast damaging weather conditions with greater reliability on the basis of recorded weather data. The results were the extraction of certain rules from the trained network, which were proposed to be able to furnish the prediction of oncoming thunderstorms, based on relevant weather parameters, in human-understandable form.

In order to allow the key stakeholders to have more float time to take appropriate precautionary and preventive measures, an accurate prediction of water quality pollution is very significant. Since a variety of existing water quality models involve exogenous input and different assumptions, artificial neural networks have the potential to be a cost-effective solution. Chau (2005) presented the application of a split-step PSO model for training perceptrons to forecast real-time algal bloom dynamics in Tolo Harbour, Hong Kong. The advantages of the global search capability of the PSO algorithm in the first step and the local fast convergence of the Levenberg–Marquardt algorithm in the second step were coupled. The results indicated that, when compared with the benchmarking backward propagation algorithm and the usual PSO algorithm, it accomplished a higher accuracy in a much shorter time.

Cheng *et al.* (2005) implemented several ANN models with a feedforward and backpropagation network structure coupled by employing a multitude of training algorithms for predicting daily and monthly river flow discharges in Manwan Reservoir. These models were compared with a conventional auto-regression time-series flow prediction model so as to test their applicability and performance. Results illustrated that the ANN models furnished better accuracy in predicting river flow than the benchmarking conventional time-series model. Amongst these models, the scaled conjugate gradient algorithm acquired the highest correlation coefficient and the smallest root mean square error. This ANN model was ultimately adopted in the real water resource project of Yunnan Power Group.

Habib and Meselhe (2006) made use of advanced computation-intensive techniques, such as neural networks and local non-parametric regression, to model highly non-linear, non-unique, and complex stage-discharge relationships for coastal low-gradient streams in south-western Louisiana. Whilst it was quite difficult for conventional methods, such as parametric regression, to model the exhibited multiple loops successfully, it was shown that both neural network and local regression models were able to reproduce the features at the outlet of the stream. Moreover, it was shown that the models had higher prediction accuracy with high flows than with low flows.

Muttil and Chau (2007) used two extensively used ML techniques, ANN and genetic programming (GP), for determining the significant input variables. The efficacy of these techniques was first illustrated on a test problem with known dependence and then they were applied to a real-life case study of water quality data in Tolo Harbour, Hong Kong. These ML techniques overcame some of the constraints of the contemporary techniques for input variable selection. The interpretation of the weights of the trained ANN and the GP evolved equations illustrated their capability of identifying the ecologically significant variables accurately. The significant variables determined by the ML techniques also showed that chlorophyll-a itself was the most significant input in forecasting the algal blooms. This indicated an auto-regressive nature or persistence in the algal bloom dynamics, which was possibly associated with the long flushing time in the semi-enclosed coastal waters. The study also concurred with the previous understanding that the algal blooms in coastal waters of Hong Kong had a typical life cycle of the order of one to two weeks.

Pape *et al.* (2007) compared the performance of several data-driven models in predicting the temporal evolution of near-shore sandbars, using daily observations of an outer sandbar at the double-barred Surfers Paradise, Gold Coast, Australia. Whilst previous results by conventional process-based models suggested that the evolution of sandbars depended non-linearly on the wave forcing, and that a time-series of sandbar positions exhibited dependencies spanning several days, their results were quite different. Their results demonstrated that non-linear effects exposed themselves for larger prediction horizons, and that there was no significant difference between non-recurrent and recurrent methods, denoting that the effects of dependencies spanning several days were of no significance.

Uddameri and Honnungar (2007) used an information-analytic technique called rough sets to understand groundwater vulnerability characteristics in 18 different counties of South Texas, in which the coastal semi-arid region was undergoing significant growth causing an enormous burden on its limited water resources. They proposed the coupling of rough sets with GIS to cluster counties exhibiting similar vulnerability characteristics and to acquire other pertinent insights. It was found that the groundwater vulnerability exhibited greater variability along the coast than in the interior sections. This might shed a new way for regional planners and environmental managers with a role in sustainable water resources management and land use development.

The determination of the longshore sediment transport rate is entailed for the planning, operation, design and maintenance of harbour and coastal engineering facilities. Singh *et al.* (2007) presented a novel method based on a combination of two soft computing tools, namely neural networks and genetic programming, as an alternative approach to the conventional empirical equations. This hybrid method was found to generate better results than the use of neural networks or genetic programming solely. It was believed that a better prediction was due to the combined effect of the ability of the neural network to approximate a non-linear function and the efficiency of the genetic programming to make an optimum search over the solution domain.

Information on tidal currents is useful in taking operation- and planning-related decisions such as the towing of vessels and monitoring of oil slick movements. Charhate *et al.* (2007) discussed a few alternative approaches based on the soft computing tools of ANNs and GP, as well as the hard mathematical approaches of stochastic and statistical methods for real-time prediction of tidal currents in the Gulf of Khambhat, India. A univariate time-series of coastal currents was employed to forecast their future values. It was found that the soft computing schemes of GP and ANN performed better than the traditional hard technique of harmonic analysis in the present application.

The significance of continuous wave data measurements in providing real-time wave information for coastal and ocean related activities and in forming a wave database useful for predicting future events using statistical or stochastic techniques are well recognized. However, sometimes the loss of data from wave buoys is inevitable. Kalra and Deo (2007) employed one of the latest soft computing tools, GP, to restore missing wave heights for six selected buoy locations along the west coast of India. The performance of GP was evaluated to be reliable in terms of the error statistics of bias, root mean square error, correlation coefficient and scatter index between the restored wave records and field observations. Londhe (2008) presented the use of soft computing techniques such as ANN and GP to retrieve the lost data by forming a network of wave buoys in the Gulf of Mexico. In order to retrieve lost data at a location, a network for each buoy was developed as the target buoy with five other input buoys. The comparison of result performance showed the superiority of GP over ANN as evident by higher correlation coefficient between observed and predicted wave heights.

Kumar *et al.* (2008) delineated the development of a comprehensive tsunami travel times atlas furnishing expected times of arrival (ETA) to various coastal destinations on the Indian Ocean rim for inclusion in the real-time tsunami warnings. They proposed the application of soft computing tools such as ANN for the prediction of the ETA in a real-time mode. They reasoned that ANN had an advantage in producing ETAs in a much faster time and also simultaneously preserving the consistency of prediction. It was suggested that modern technology could prevent or help minimize the loss of life and property.

An accurate and timely prediction of river flow flooding can provide time for the authorities to take pertinent flood-protection measures such as evacuation. Various data-derived models including LR (linear regression), NNM (the nearest-neighbour method) ANN and SVR (support vector regression) have been successfully applied to water level prediction. Of these, SVR is particularly highly valued, because it has the advantage over many data-derived models of overcoming overfitting of training data. However, SVR is computationally time-consuming when used to solve large-size problems. In the context of river flow prediction, equipped with an LR model as a benchmark and genetic algorithm-based ANN (ANN-GA) and NNM as counterparts, Wu *et al.* (2008) proposed a novel distributed SVR (D-SVR) model. It implemented a local approximation to training data because partitioned original training data was independently fitted by each local SVR model. ANN-GA and LR models were also used to help determine input variables. A two-step GA algorithm was employed to find the optimal triplets (C, ε, σ) for the D-SVR model. The validation results revealed that the proposed D-SVR model could carry out the river flow prediction better in comparison with others, and dramatically reduced the training time compared with the conventional SVR model. The pivotal factor contributing to the performance of D-SVR might be that it implemented a local approximation method and the principle of structural risk minimization.

One of the important issues in coastal and offshore engineering is wave parameters prediction. Mahjoobi *et al.* (2008) presented alternative models based on ANNs, a fuzzy inference system (FIS) and an adaptive-network-based fuzzy inference system (ANFIS) to hindcast the wave parameters (significant wave height, peak spectral period and mean wave direction) in Lake Ontario. Wind speed, wind direction, fetch length and wind duration were used as input variables. Result comparisons indicated that the error statistics of various soft computing models were similar. Mahjoobi and Adeli Mosabbeb (2009) employed SVM, a strong machine learning and data mining tool, to predict significant wave height in Lake Michigan. Field observations of current wind speed as well as data for up to six previous hours were used as input variables. Comparisons indicated that the error statistics of the SVM model marginally outperformed conventional ANN, simultaneously requiring much less computational time.

6.4 Knowledge-based expert systems

When compared with other AI techniques, KBS is quite distinct and has apparent advantages over the others in allowing more transparent transfers of knowledge in the application and interpretation of the model, offering expert support to novice users, and achieving effective communication from developers to users.

Ranga Rao and Sundaravadivelu (1999) presented a knowledge-based expert system, KNOWBESTD, employing a shell LEVEL5 OBJECT for the design of berthing structures such as quays, wharfs, piers, jetties and dolphins. The most economical design of a typical berthing structure, accomplished through the assistance of this expert system, was demonstrated. These structures would be checked against the limit state of cracking in order to attain an important goal of corrosion prevention. The usefulness of this system was that since construction and maintenance of these structures were very expensive, significant monetary savings could be achieved when an optimal as well as sustainable design was adopted.

Moore *et al.* (1999) described the development of a coastal management expert system and the application of the system to characterize beach morphology on the rapidly eroding Holderness coast, eastern England. It was a decision-making support tool that applied expert knowledge to help the coastal zone manager in monitoring and managing coasts with long-term erosion problems. The constituent features of a composite ridge-type landform were elicited and stored as expert knowledge or rules, in terms of positional relationships and morphometric parameters including slope, aspect and convexity. These rules were employed on consecutive digital elevation models to extract a geomorphological feature.

Chau *et al.* (2002) addressed a prototype knowledge management system on flow and water quality to mimic human expertise during the problem-solving by integrating artificial intelligence with various types of descriptive knowledge, procedural knowledge, and reasoning knowledge in the coastal hydraulic and transport processes. The system was developed through utilizing Visual Rule Studio, a hybrid expert system shell, as an ActiveX Designer under the Microsoft Visual Basic 6.0 environment, which coupled the advantages of both production rules and object-oriented programming technology. The architecture, the development and the implementation of the prototype system were described in detail. In accordance with the succinct features and conditions of a variety of flow and water quality models, three kinds of class definitions, Section, Problem and Question, were categorized and the corresponding knowledge rule sets were also set up. During the inference process, both forward chaining and backward chaining were employed collectively. Application of the prototype knowledge management system was illustrated by employing a typical example case.

Anuchiracheeva *et al.* (2003) presented the method of effective systematization, analysis and visual display of local knowledge using a GIS for use in fisheries management at Bang Saphan Bay, Thailand, in order to facilitate its use by policymakers. Field observations of location fished, time of fishing, techniques and technology used, and species targeted were acquired from local fishers and then mapped using GIS. Thus, local fisheries knowledge could be converted into geo-spatial data form, and the succinct results could be employed easily to guide fishery management and planning.

Fdez-Riverola and Corchado (2003) developed a hybrid neuro-symbolic problem-solving model in order to forecast parameters of a complex and dynamic environment, such as the prediction of the red tides appearing in the coastal waters of the north-west of the Iberian peninsula. This model attempted to solve the difficult problems of predicting the parameter values that determined the characteristic behaviour of a system when the governing equations of that system were unknown. The system employed a case-based reasoning model to couple a growing cell structures network, a radial basis function network and a set of Sugeno fuzzy models to retrieve, adapt and review the proposed solution and, all in all, to furnish an accurate prediction.

Dai *et al.* (2004) delineated a knowledge base for watershed assessment for sediment (WAS), which was tailored for protection of the fish habitat and control of excessive sediment, and was applied as a decision support tool to evaluate the condition of a coastal watershed in northern California, United States. The WAS model furnished a means to assemble key pieces of information and reasoning that support land use or regulatory decisions, and to communicate among diverse audiences the basis for those decisions. In this way, experts from diverse fields could contribute to an integrated assessment of watershed conditions, which was a complex problem often with ill-defined issues and lacking data.

Chau (2006b) presented the integration of the recent advances in AI technology with contemporary numerical models to form an integrated KBS on flow and water quality. A hybrid application of the latest AI technologies, namely, KBS, artificial neural networks and fuzzy inference systems, was adopted for this domain problem. This prototype system could act as both a design aid and a training tool, thus enabling hydraulic engineers and environmental engineers to become acquainted with up-to-date flow and water quality simulation tools. More importantly, it could fill the existing gaps between researchers and practitioners in the application of recent technology in addressing real prototype problems in Hong Kong. This integrated system could quickly help policymakers arriving at decisions and offer a convenient and open information service on water quality for the general public.

Schories *et al.* (2009) developed a classification approach within the European Water Framework Directive for the outer coastal waters of the German Baltic Sea, with focus on the known recent presence and depth distribution of two specific species of plant. The boundaries of the ecological status according to the Water Framework Directive were computed based on modelling. This model allowed the adaption of the boundaries calculations to new knowledge about historical data and the ecophysiological light demand of plants.

Pereira and Ebecken (2009) employed a machine-learning approach to determine the ecological status of coastal waters at Cabo Frio Island in Rio de Janeiro, based on patterns of the occurrence of fauna as well as its relationship with other environmental parameters. This location has been suffering from anthropogenic impact. Models of crisp and fuzzy rules were tested as classifiers. Results indicated that it was possible to access hidden patterns of water masses within a set of association rules, which might be useful for decision-making, system management and sustainable management of marine resources.

6.5   Manipulation of conventional models

During the past decade, the general availability of sophisticated personal computers with ever-expanding capabilities has given rise to increasing complexity in terms of computational ability in the storage, retrieval, and manipulation of information flows. With the recent advances in AI technology, there has been an increasing demand for a more integrated approach in addition to the need for better models. Justification for this claim comes from the relatively low utilization of models in the industry when compared to the number of reported and improved models. It is expected that this enhanced capability will both increase the value of the decision-making tool to users and expedite the water resources planning and control process.

Chau and Chen (2001) presented a fifth generation numerical modelling system in coastal zones, by employing the recent advances in AI technologies. The expert system technology was coupled into the modelling system for coastal water processes with conventional numerical computational tools, data and graphical preprocessing and postprocessing techniques. Five kinds of knowledge bases were established in the system to delineate the existing expert knowledge about model parameters, relations between parameters and physical conditions, and various possible selections for parameters and rules of inference. The inference engine was tailored to be driven by the confidence of correctness, and the rule base was constructed with the factor of confidence to link different relations. The decision tree was such as to drive the inference engine to explore the route of the selection procedure on modelling. The decision tree relied upon the real problem specifications and could be adjusted during the dialogue between the system and the user. The forward chaining and backward chaining inference techniques were mixed together in the system to help matching the parameters in the model and the possible selections with confidence higher than the threshold value. The expert system technology was successfully incorporated into the system to furnish assistance for model parameter selection or model selection, and to render the numerical model system more accessible for non-expert users.

It was noted that the outcomes of the algorithmic execution of numerical flow and water quality models often differed from those expected, in particular when the model was built initially. This required the modeller to undertake the manipulation procedure, which comprised feedback and modification. Hence, it was desirable that expert system technology be incorporated into the modelling system to offer help for the novice user who lacked the required knowledge to set up the model and evaluate the results. Chau (2003) developed and implemented a prototype expert system on the manipulation of numerical coastal flow and water quality models by using an expert system shell. It was demonstrated that, through the successful development of this prototype system, the expert system technology could be integrated into numerical modelling for mimicking the manipulation process. It helped the user to set up an appropriate strategy for arriving at a balance between accuracy and effectiveness and to tune the model to attain successful simulation of real phenomena. It was capable of bridging the existing gap between numerical modellers and practitioners in this field.

Currently, the numerical simulation of flow and/or water quality is becoming more and more sophisticated. There arises a demand on the integration of recent knowledge management (KM) and artificial intelligence technology with conventional hydraulic algorithmic models in order to assist novice application users in the selection and manipulation of various mathematical tools. Chau (2007) proposed an ontology-based KM system (KMS), which used a three-stage life cycle for the ontology design and a Java/XML-based scheme for automatically generating knowledge search components. The prototype KMS on flow and water quality was developed to mimic human expertise during the problem-solving by integrating artificial intelligence with various knowledge involved in the coastal hydraulic and transport processes. The ontology was categorized into information ontology and domain ontology so as to realize the objective of a semantic match for knowledge search. The application of the prototype KMS was illustrated through a case study.

Sheng and Kim (2009) evaluated the predictive skills of an integrated physical-biogeochemical modelling system for shallow estuarine and coastal ecosystems in the Indian River Lagoon estuarine system. Model skills for hydrodynamic and water quality simulations were assessed in terms of the absolute relative errors and the relative operating characteristic scores. Both methods illustrated that the modelling system had skills in simulating water level, salinity, dissolved oxygen, chlorophyll, dissolved nutrients, etc.

Moreover, results suggested that model skills could be improved with more detailed sediment-water quality data, addition of a coastal ocean domain, and improved knowledge of model parameters/coefficients.

6.6 Conclusions

A general introduction to some contemporary soft computing techniques is given in this chapter. Out of the various possible ML techniques, we considered ANN, fuzzy logic and evolutionary algorithms to represent distinct attributes. ANN is the most widely used method in water resources variable modelling (Maier and Dandy 2000). Fuzzy logic and evolutionary algorithms have an advantage in their ability to generate equations or formulae relating input and output variables, which might provide physical insight into the ecological processes involved. Further, Recknagel (2001) has reported that ANN and genetic algorithms currently appear to be the most innovative for ecological modelling. Knowledge-based expert systems were highlighted for their ability to transfer knowledge in the application and interpretation of models, to offer expert support to novice users, and to achieve effective communication from developers to users. We present more details of these techniques one by one in the following chapters.

7 Artificial neural networks

7.1 Introduction

An artificial neural network (ANN) is a computing paradigm tailored to mimic natural neural networks (Haykin 1999). It can be defined as "a computational mechanism able to acquire, represent, and compute a mapping from one multivariate space of information to another, given a set of data representing that mapping" (Garrett 1994). A typical ANN comprises an input layer that receives inputs from the environment, an output layer that generates the network's response, and some intermediate hidden layers. Maier and Dandy (2000) furnished a comprehensive review on using neural network models to predict and forecast water resources variables. The basis of ANNs is our current understanding of the brain and its pertinent nervous systems. Numerous processing elements connected by links of variable weights are grouped together to constitute black box representations of systems. In this chapter, the characteristics of ANNs and the commonly used backpropagation forward-feeding ANN are delineated. Two real applications of ANN are also demonstrated. The first application case study presents the analysis of algal dynamics data from a coastal monitoring station. The second application is for prediction of long-term flow discharges in Manwan based on historical records.

7.2 Supervised learning algorithm

An ANN comprises typically several layers of interconnected neurons, each of which is in turn linked to other neurons in the following layer. Data are fed to the neural network through an input layer. The output layer then keeps the response of the network to the input. In many cases, a certain number of hidden layers may exist between the input layer and the output layer. The mechanism is such that all hidden and output neurons process their inputs by multiplying each input by its weight, adding their product, and then applying a non-linear transfer function to the sum to produce a resulting output. It is one of the data-driven models and has the ability to learn complex model functions from examples after appropriate training (Rumelhart *et al.* 1994).

Training can be defined as the process of adjusting the connection weights in the neural network in order to attach the best matches between the network's response and the targeted response. Whilst this step can be addressed by employing an optimization method, the backpropagation method avoids this costly procedure by employing an approximation to a gradient descent method. In the forward pass, each neuron computes a response from the weighted sum of its inputs from neurons linked to it, employing a preset activation function. The output in a layer acts as one of the inputs of other neurons in the next layer.

ANNs have found applications in prediction of water quality variables such as algal concentrations (Chau 2005), cyanobacterial concentrations (Maier and Dandy 1997), ecological modelling (Lek and Guegan 1999), phosphorus (Zou *et al.* 2002), salinity levels (Bastarache *et al.* 1997), and so on. Kralisch *et al.* (2003) used an ANN approach to attain optimization and balance of watershed management in order to compromise between water quality demands and the consequent restrictions for the farming industry. Maier *et al.* (2004) employed ANNs to forecast treated water quality parameters as well as optimal alum doses.

Recknagel *et al.* (1997) developed an ANN model for prediction of algal blooms in four different freshwater systems and compared its performance against several benchmarking conventional approaches. The water-specific limnological time-series comprised biomass of the ten dominating algae species as observed for 12 years and the measured environmental driving variables.

Scardi (2001) presented and discussed some approaches that could enhance neural network models to overcome the problem imposed by the limited amount of available data in phytoplankton primary production modelling in saltwater eutrophic areas. An adopted approach that could accomplish this purpose was to select additional inputs from a broader range of variables as copredictors. Moreover, information acquired from existing models could be effectively exploited and embedded into the ANN models by a metamodelling approach via a constrained training procedure.

Zou *et al.* (2002) employed a three-layer feed-forward backpropagation ANN to mimic the relationship between the parameters and the steady-state response of a mechanistic total phosphorus model in the Triadelphia Reservoir spanning three years of data. Figure 7.1 shows the architecture of this typical ANN. There were three nodes in the input layer, namely, settling velocity, recycling velocity and burial velocity. There were six hidden nodes in the hidden layer, and the concentration of phosphorus was the only node in the output layer. It was illustrated that, in this case study, the ANN technique has the ability to accurately approximate the input–output response of a water quality model. For both the training and testing sets,

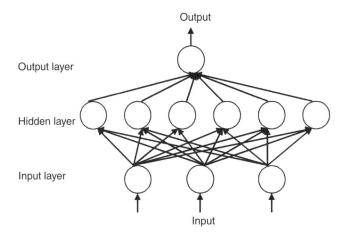

Figure 7.1 Architecture of a three-layer feed-forward backpropagation ANN

the prediction results of phosphorus concentration by the ANN agree with those generated by the benchmarking mechanistic model.

Cheng *et al.* (2005) implemented several ANN models with a feed-forward and backpropagation network structure coupled by employing a multitude of training algorithms for predicting daily and monthly river flow discharges in Manwan Reservoir. These models were compared with a conventional auto-regression time-series flow prediction model so as to test their applicability and performance. Results illustrated that the ANN models furnished better accuracy in predicting river flow than the benchmarking conventional time-series model. Amongst these models, the scaled conjugate gradient algorithm acquired the highest correlation coefficient and the smallest root mean square error. This ANN model was ultimately adopted in the real water resource project of Yunnan Power Group.

Londhe (2008) presented the use of soft computing techniques like ANN and GP to retrieve lost data by forming a network of wave buoys in the Gulf of Mexico. In order to retrieve lost data at a location, a network for each buoy was developed as the target buoy with five other input buoys. The comparison of result performance showed the superiority of GP over ANN, as evidenced by a higher correlation coefficient between observed and predicted wave heights.

It is observed that most of the studies were performed for limnological systems (Yabunaka *et al.* 1997; Recknagel *et al.* 1998; Karul *et al.* 2000) or riverine systems (Whitehead *et al.* 1997; Maier *et al.* 1998; Cheng *et al.* 2005), while literature on ANN modelling of coastal systems was relatively scarce (Barciela *et al.* 1999). It is noted that, in many studies, the effectiveness of ANN as a predictive tool might not be fully addressed. For example, the water quality dynamics at the current moment were often related via the model to other environmental variables simultaneously, which rendered them not too useful for real predictions. Moreover, most studies used almost all available environmental parameters as input variables without accounting for the optimal choice among them.

In fact, there are many options during the process of setting up an ANN model, namely, the choice of appropriate model inputs and network architecture, the division and preprocessing of the input data, the optimization of the connection weights, the adoption of performance criteria, and the validation of the model (Maier and Dandy 2000). Each of the above parameters can have a significant effect on the accuracy of a prediction.

7.3 Backpropagation neural networks

An ANN is one of the various available contemporary forms of artificial intelligence, which simulate the functioning of the human brain and nervous system. It obtains knowledge via a learning process involving the determination of an optimal set of connection weights and threshold values for the neurons. The capability of "training" and "learning" the output with the provision of a given input renders ANNs once to address large-scale, arbitrarily complex, non-linear problems (Sivakumar *et al.* 2002). Though there are numerous types of ANNs, the feed-forward neural network or the multilayer perceptron, which is organized as layers of computing elements or neurons linked via weights between layers, is nowadays the most popularly used.

A typical neural network comprises several layers. Figure 7.1 shows that the basic structure of a network is usually composed of three types of layers: the input layer, which receives inputs from the environment; the hidden layer or layers; and the output layer, which outputs the network's response to the corresponding input. The architecture of an ANN is determined by the following parameters: weights, which link neurons; a transfer function that controls the production of output in a neuron; and learning rules. Each neuron computes a response from the weighted sum of its inputs and bias (threshold value) from neurons linked to it, employing a preset activation or transfer function, which is often an S-shaped sigmoid function or a hyperbolic-tangent functions. The sigmoid function is characterized by the bounds of zero and one as the lower and upper limits, respectively, monotonically increasing, and continuous and differentiable within the entire domain. The logistic function shown as follows is one of the most commonly used sigmoid functions for ANNs:

$$f(x) = \frac{1}{1 + \exp(-x)} \tag{7.1}$$

where x is in the range $[-\infty, +\infty]$.

The feed-forward ANN is organized as layers of neurons linked by weighted connections between layers. One of the most common learning rules for the feed-forward network is the backpropagation algorithm, and a neural network with such a type of learning algorithms is termed a back-propagation network (BPN). Its training is composed of two key processes, namely, forward pass and backward pass. During the forward pass, the input data are multiplied by the initial weights. Simple summation of the weighted inputs then yields the net to each neuron. The output of a neuron is acquired when the activation or transfer function is applied to the net of a neuron. The output of the neuron, when it is transmitted to the next layer, becomes an input. The above procedure is then repeated until the output layer is arrived at. Whilst the neuron response is computed from the weighed sum of its inputs and bias with a predetermined activation function in the forward pass, the weights are adjusted based on the error between the computed and target outputs in the backward pass. The error is then dis-tributed to neurons in each layer by the derivatives of the objective function with respect to the weights, which can be moved in the direction in which the error declines most quickly by using a gradient descent method.

Mathematically, the representation is as follows:

$$H_j = \sum_{i=1}^{k} w_{ij} x_i + \theta_j, \ j = 1, \ldots, h \qquad (7.2)$$

where H_j is the weighted sum of outputs of the jth hidden node from the previous layer, w_{ij} is the connection weight from the ith input neuron to the jth hidden neuron, x_i is the input value, k is the number of input nodes, θ_j is a threshold or bias, and h is the number of hidden nodes. Each hidden node is then transformed via a sigmoid function to generate a hidden node output HO_j as follows:

$$HO_j = f(H_j) = \frac{1}{1 + \exp[1 - (H_j + \theta_j)]} \qquad (7.3)$$

Similarly, at the output layer, the following equation can be written:

$$IO_n = \sum_{j=1}^{h} w_{jn} HO_j, \ n = 1, \ldots, m \qquad (7.4)$$

where IO_n is the weighted sum of outputs of the nth output node from the previous hidden layer, w_{jn} is the connection weight from the jth hidden neuron to the nth output neuron, and m is the number of output nodes. The neural output value O_n is then obtained by applying the sigmoidal function to IO_n.

For backpropagation networks, the derivative of the activation function is employed to effect adjustment of the network weights. Hence, at the end of each forward pass, the error between the computed outputs of the network and the target outputs is computed.

The mean square error E for all input patterns is shown as follows:

$$E = \frac{1}{2N} \sum_{p=1}^{N} \sum_{n=1}^{m} (T_{pn} - O_{pn})^2 \tag{7.5}$$

where N is the number of data input patterns, T_{pn} is the target value for the pth pattern, and O_{pn} is the neural network output value for the pth pattern.

The termination criterion is reached when the error is smaller than a preset value. If the error is larger than a predetermined value, the procedure continues with a backward pass; otherwise, the training is stopped. During the backward pass, the weights in the network are adjusted by employing the error value. It should be mentioned that the modification of weights in the output layer is usually different from the hidden layers. This is because, in the output layer, the target outputs are given whilst in the intermediate hidden layers, there are no target values. Hence, backpropagation employs the derivatives of the objective or fitting function with respect to the weights in the entire network to distribute the error to neurons in each layer of the network (Tokar and Johnson 1999). A gradient descent method is often used which moves the weights in a direction in which the error reduces in a quicker manner:

$$\delta_n = O_n (1 - O_n) (T_n - O_n) \tag{7.6}$$

where δ_n is the gradient for each neuron on the output layer, T_n is the correct teaching value for the output unit n, and O_n is the neural network output. The error gradient δ_j is then recursively computed for the hidden layers as follows:

$$\delta_j = HO_j(1 - HO_j) \sum_{n=1}^{m} \delta_n w_{jn} \tag{7.7}$$

The errors are propagated backwards until the input layer. The error gradients are then employed to update the network weights as follows:

$$\Delta w_{ji}(r) = \eta \delta_j x_i \tag{7.8}$$

$$\Delta w_{ji}(r + 1) = w_{ji}(r) + \Delta w_{ji}(r) \tag{7.9}$$

where r is the iteration number, and η is the learning rate which furnishes the step size during the gradient descent. Whilst a larger value of learning rate can speed up the convergence process, it also results in oscillations in the solution. Thus, in choosing an appropriate learning rate, a compromise has to be made between reliability of results and speed of training time.

Another commonly used form of the weight change, allowing the use of a larger value of learning rate, is as follows:

$$\Delta w_{ji}(r) = \eta \delta_j x_i + \alpha \Delta w_{ji}(r-1) \qquad (7.10)$$

where α is the momentum coefficient employed to enhance the convergence. In the above equation, the momentum coefficient effectively makes the weight change at iteration r proportional to the previous weight change at iteration $r-1$, thus reducing the possibility of oscillations.

Though an ANN can be considered a flexible and powerful mapping tool, an adequate initialization of weights and biases has a significant influence on the performance of the network, and an inappropriate assignment of weights and biases can result in local convergence. More in-depth discussions on ANNs are found in Rumelhart *et al.* (1986), Simpson (1990), and Haykin (1999).

7.4   Advantages and disadvantages of artificial neural networks

The major advantage of ANNs over other modelling techniques is their capability to simulate complex and non-linear processes without having to determine the exact form of the relationship between input and output variables. The learning process in ANNs involves only the adjustment and modification of the connection weights amongst neurons in different layers. Pattern matching, combinatorial optimization, data compression, and function optimization are contemporary application areas and topics that have been addressed by ANN techniques. ANNs, being a developing and promising technique nowadays, have become very popular for both prediction and forecasting in different fields. Their capability to address uncertainty in complex situations for wide-ranging application domains is proven from the literature in recent years (Serodes and Rodriguez 1996; Whitehead *et al.* 1997; Maier *et al.* 1998; Chau and Cheng 2002; Cheng *et al.* 2005).

It should be noted that the initialization of weights and biases may also have some effects on the network performance, and hence improper assigned values can result in local convergence. The major drawback of the conventional BPN with a gradient descent learning algorithm is the slow convergence rate. Thus, among others, although the steepest descent method is the simplest and most popular, it lacks effectiveness owing to slow convergence and its vulnerability to getting stuck in a local minimum. In real ANN applications, the steepest descent method is seldom used. Different algorithms have been developed to overcome this drawback. Haykin (1999) discussed several data-driven optimization training algorithms such as the Levenberg–Marquardt (LM) algorithm and the scaled conjugate gradient (SCG) algorithm.

Another point that needs attention is that ANNs cannot extrapolate beyond the range of the training data. Hence they may not be capable of controlling for data trends or heteroscedasticity. If ANNs are to take a significant role in coastal modelling, careful selection of various input and output parameters at different steps as well as a thorough comprehension of the boundaries of applicability are required. Moreover, an ANN is currently considered as a black box tool only. More attention should be paid to extracting some knowledge from the learning process. More effort can be given to the application of this technique to coastal modelling in the future.

Wu *et al.* (2009) endeavoured to couple three data-preprocessing techniques – moving average (MA), singular spectrum analysis (SSA), and wavelet multi-resolution analysis (WMRA) – with an ANN in order to enhance the estimate of daily flows. Six models, including the original ANN model without data preprocessing – ANN-MA, ANN-SSA1, ANN-SSA2, ANN-WMRA1, and ANN-WMRA2 – were developed and assessed. The ANN-MA, ANN-SSA1, ANN-SSA2, ANN-WMRA1 and ANN-WMRA2 were developed by employing the original ANN model coupled with MA, SSA and WMRA, respectively. Two different means were used for SSA and WMRA. The models were applied to two daily flow series in two watersheds in China, Lushui and Daning, for three different prediction horizons, namely, one-, two-, and three-day-ahead forecasting. Results indicated that, among the six models, the ANN-MA has the highest accuracy and is able to eradicate the lag effect. It was also noted that the performances from the different means used for SSA and WMRA did not affect the results. Moreover, in that case study, the models based on the SSA performed better than their counterparts of the WMRA at all forecasting horizons. It indicated that the SSA was more effective than the WMRA in enhancing the ANN performance.

7.5 Prototype application I: algal bloom prediction

The first prototype application of ANN is for real-time algal bloom prediction at Tolo Harbour in the north-eastern coastal waters of Hong Kong. The nature of the data employed and details of the modelling are shown in the following sections.

7.5.1 Description of the study site

Figure 7.2 shows the location of Tolo Harbour, which is a semi-enclosed bay in the northeastern coastal waters of Hong Kong and is connected to the open sea via Mirs Bay. It is generally recognized that the water quality gradually deteriorates from the better flushed outer "Channel Subzone" towards the more enclosed and densely populated inner "Harbour Subzone".

Over the past few decades, the nutrient enrichment exhibited by the eutrophication phenomenon in the harbour resulting from municipal and

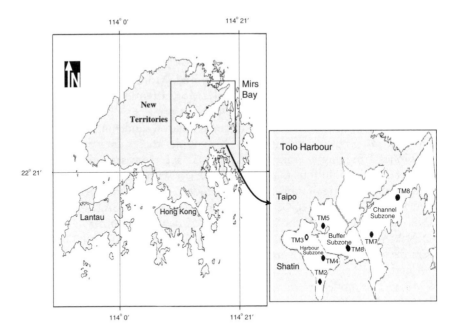

Figure 7.2 Location of Tolo Harbour

livestock waste discharges has long been a major environmental concern and threat. Two major treatment plants at Shatin and Taipo are continuously releasing organic loads. There are also non-point sources from runoff and direct rainfall, as well as waste from mariculture. The eutrophication effect has generated frequent algal blooms and red tides, which occurred particularly in the weakly flushed tidal inlets inshore. As a consequence, occasional massive fish kills are caused owing to severe dissolved oxygen depletion or toxic red tides. Numerous studies have been made, which concluded that the ecosystem health state of the Tolo Harbour had been progressively deteriorating since the early 1970s (Morton 1988, Xu *et al.* 2004). During this period, serious stresses to the marine coastal ecosystem were caused by the nutrient enrichment in the harbour due to urbanization, industrialization, livestock rearing, etc. Frequent occurrences of red tides and associated fish kills were recorded in the late 1980s, which was possibly the worst period. Morton (1988) considered that the Tolo Harbour was "Hong Kong's First Marine Disaster" and that the inner harbour was effectively dead. At that time, since Tolo Harbour had entered a critical stage, the Hong Kong Government decided to develop and implement an integrated action plan, the Tolo Harbour Action Plan (THAP). Through the implementation of THAP in 1988, significant effectiveness in the reduction of pollutant loading and in the enhancement of the water quality was recorded.

During the past few decades, a number of field and process-based modelling studies on eutrophication and dissolved oxygen dynamics of this harbour have been reported (e.g., Chan and Hodgkiss 1987; Chau and Jin 1998; Lee and Arega 1999; Chau, 2004a; Xu *et al.* 2004).

In this study, an ANN model for real-time algal bloom prediction at Tolo Harbour is developed and implemented. The data used in this study are mainly the monthly/biweekly water quality data collected by the Hong Kong Government's Environmental Protection Department as part of its routine water quality monitoring programme. The data from the most weakly flushed monitoring station, TM3 (location shown in Figure 7.2), are chosen so as to separate the ecological process from the hydrodynamic effects as far as possible. All data of ecological variables are collected and presented as depth-averaged. In order to obtain the daily values, linear interpolation of the biweekly observed data is performed. Moreover, some meteorological data including wind speed, solar radiation and rainfall recorded by the Hong Kong Observatory are employed. No interpolation is thus required on these sets of data since they are recorded in a daily manner. The data are split into two portions: data between 1988 and 1992 for training, and data between 1993 and 1996 for testing the models. A more detailed description of the water quality data in this study site can be found in Lee and Arega (1999) and Lee *et al.* (2003).

7.5.2 Criterion of model performance

The evaluation of the performance of the predictions for ANN is made by two goodness-of-fit measures: the root mean square error (RMSE) and the correlation coefficient (CC). Visual comparison can be made on time-series plots.

$$\text{RMSE} = \sqrt{\frac{\sum_{i=1}^{p}\left[(X_m)_i - (X_s)_i\right]^2}{p}} \tag{7.11}$$

$$\text{CC} = \frac{\sum_{i=1}^{p}\left[(X_m)_i - \left(\overline{X}_m\right)_i\right]\left[(X_s)_i - \left(\overline{X}_s\right)_i\right]}{\sqrt{\sum_{i=1}^{p}\left[(X_m)_i - \left(\overline{X}_m\right)_i\right]^2\left[(X_s)_i - \left(\overline{X}_s\right)_i\right]^2}} \tag{7.12}$$

where subscripts m and s = the measured and simulated chlorophyll-a levels, respectively; p = total number of data pairs considered; and $\overline{X}_m$ and $\overline{X}_s$ = mean value of the measured and simulated data, respectively. RMSE furnishes a quantitative indication of the model error in units of the variable, with the attribute that larger errors draw greater attention than smaller ones. The coefficient of correlation between the measured and simulated data can be considered a qualitative evaluation of the model performance.

7.5.3 Model inputs and output

The model comprises nine input variables and one output variable. Nine input variables are chosen to be the most influential factors on the algal dynamics of Tolo Harbour:

1. chlorophyll-a, Chl-a (μg/L);
2. total inorganic nitrogen, TIN (mg/L);
3. dissolved oxygen, DO (mg/L);
4. phosphorus, PO_4 (mg/L);
5. secchi-disc depth, SD (m);
6. water temperature, Temp ($^\circ$C);
7. daily rainfall, Rain (mm);
8. daily solar radiation, SR (MJ/m^2); and
9. daily average wind speed, WS (m/s).

These are determined after taking careful consideration of previous computer modelling and field observation studies in the weakly flushed embayment in Tolo Harbour (Chau *et al.* 1996; Lee and Arega, 1999; Lee *at al.* 2003). The sole model output is chlorophyll-a, which serves as a good indicator of the algal biomass. The model is applied to attain one-week-ahead prediction of algal blooms. This time period is selected after having taken considerations of the ecological process at Tolo Harbour and practical constraints of data collection. A time lag of 7–13 days is introduced for each of the input variables. It should be noted that, with a targeted one-week lead-time of the prediction, the time lag has to start from 7.

7.5.4 Significant input variables

The selection of the appropriate model input variables is very significant for any machine learning (ML) technique. As shown in the last section, the selection of input variables is in general according to a priori knowledge of causal variables and physical/ecological insight of the modeller into the problem domain. Moreover, the use of lagged input variables results in better predictions in a complex and dynamical system. Maier and Dandy (2000) have made a thorough review of 43 international journal papers published between 1992 and 1998, which employed ANNs for simulation and prediction of water resources variables. They found that the modelling process is not performed in a correct manner in many articles, and that arbitrary selection of model inputs is an area of great concern.

In this model, ANN networks are trained to find an empirical relationship between the nine selected input variables with time lag of 7–13 days and the chlorophyll-a concentration at time t. For each of the nine input variables, there are seven time-lagged variables, rendering a total of 63 ($= 9 \times 7$) input variables. In order to reduce the computational effort, there is a necessity to find out the significant input variables, out of these 63 variables.

A fully connected feed-forward multiple layer perceptron (MLP) neural network trained with a backpropagation algorithm with momentum term was employed for the prediction of algal blooms. A single hidden layer was adopted. Hence, the resulting MLP neural network structure comprises three layers: an input layer, a hidden layer and an output layer.

There are 63 input nodes in the input layer, corresponding to the input variables determined in the last section. Great care has to be exercised in the determination of the optimal number of nodes in the hidden layer, which significantly affects the performance of the trained network. It is generally recognized that networks with fewer hidden nodes are preferred, owing to better generalization capability and less over-fitting problem. Moreover, with smaller numbers of nodes, less computational effort is accomplished. Nevertheless, a balance has to be struck. If the number of nodes is too small, it might not be able to capture the underlying behaviour of the data, and the performance of the network will deteriorate. In order to determine the optimal number of nodes in the hidden layers, a trial and error procedure was undertaken by gradually varying the number from 2 to 10. Ultimately, the optimal number of hidden nodes is found to be six. The output layer of the networks comprises only one neuron, which is the chlorophyll-a concentration to be predicted at time t.

The trial and error method is also used to determine the learning rate and the momentum coefficient. For this neural network run, the optimal values for the learning rate and the momentum coefficient are found to be 0.05 and 0.1, respectively. The hyperbolic-tangent function (tanh) is chosen as the transfer function for both the hidden and the output layers.

Figure 7.3 shows the network structure employed to forecast the algal biomass with a one-week lead-time. Based on the model formulation and training details as described above, simulations are then performed. It is noted that, after 500 epochs in all the simulations, the backpropagation training is stopped. Again, this number was adopted by trial and error. It can be observed that 500 epochs are good enough to train the network without any over-training.

An analysis of the network weights is undertaken to select the significant input variables in this case study. It is noted that, in the trained network, the connection weights along the paths from the input nodes to the hidden nodes indicate the relative predictive significance of the independent variables. There are a total of 63 input nodes in this neural network. A new term, the contribution factor, is devised to measure the significance of any input variable in forecasting the network's output, relative to the remaining variables within the same network. The definition of the contribution factor

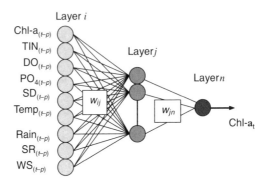

Figure 7.3 ANN model for the forecasting of algal blooms ($p = 7 \ldots 13$ for one-week forecast; $p = 14 \ldots 20$ for biweekly forecast)

of the ith variable, CF_i, is as follows:

$$CF_i = \frac{\sum_{j=1}^{h} ABS\left(w_{ij}\right)}{\sum_{i=1}^{k}\sum_{j=1}^{h} ABS\left(w_{ij}\right)} \times 100 \tag{7.13}$$

where h is the number of hidden nodes, k is the number of input variables, w_{ij} are the weights from input layer i to the hidden layer j (see Figure 7.3), and *ABS* denotes the absolute function.

Table 7.1 lists the contribution factor of each of the 63 input variables computed by employing equation (7.13). As can be observed in the table, the sum of the contribution factors of all the 63 input variables should add up to 100 per cent. The definition of the contribution factor is designed such that a higher value for a particular variable denotes the greater contribution exerted by that variable on the prediction, mainly based on the weights of the trained neural network. It can be observed from this analysis that Chl-a at $(t-7)$, having a contribution factor of 9.17 per cent, is the most significant constituent in predicting the one-week-ahead algal biomass. Other variables are PO_4, TIN, DO and SD, having a contribution factor larger than 2.00 per cent. The variables considered to be more significant are shaded in Table 7.1.

From Table 7.1, it is apparent that, apart from Chl-a at $(t-7)$ being the most significant in predicting itself, all Chl-a values during the preceding one to two weeks are also significant. This indicates an auto-regressive nature or "persistence" for the algal dynamics, which is common for geophysical time-series. Such an auto-regressive nature is frequently exhibited by these types of time-series owing to their inherent inertia or carryover process in the physical system. Numerical modelling can contribute to comprehension

Table 7.1 Contribution factors on the basis of the weights of the trained ANN for one-week predictions

Input variables	Contribution factors of the input variables (%)*							Sum
	(t−7)	*(t−8)*	*(t−9)*	*(t−10)*	*(t−11)*	*(t−12)*	*(t−13)*	
Chl-a	9.17	2.62	3.35	2.27	2.18	1.89	3.05	24.53
TIN	3.37	1.54	1.18	1.68	1.50	1.23	2.27	12.77
DO	3.12	2.06	1.37	1.41	1.13	1.05	1.55	11.68
PO_4	4.11	1.62	1.04	0.72	1.04	1.69	2.55	12.77
SD	1.08	1.20	1.11	1.22	1.37	1.70	2.98	10.66
Temp	1.42	1.01	1.28	1.00	0.78	1.02	1.70	8.20
Rain	0.95	1.31	0.97	1.08	1.15	1.18	1.31	7.95
SR	1.00	0.56	0.69	0.65	0.68	0.42	0.69	4.70
WS	1.09	0.89	1.06	1.05	1.13	0.73	0.79	6.74

Sum of contribution factors of all variables = 100

* Shaded variables have a contribution factor greater than 2%.

of the physical system by revealing factors about the process that establishes persistence into the series. In this case study, the long flushing time or residence time in the semi-enclosed coastal waters might be the key factor causing the auto-regressive nature of chlorophyll dynamics as revealed by the ML techniques. It should be pointed out that, in the inner and outer Channel Subzones, the tidal currents are very small, with the average current velocity being 0.04 m/s and 0.08 m/s, respectively (EPD 1999). The weak tidal flushing in the harbour is therefore due to its landlocked nature. Lee and Arega (1999) suggested that the flushing times in the inner Harbour Subzone are on the order of one month.

Results indicate that the nutrients, i.e. PO_4 and TIN, together with DO and SD to a lesser extent, are also significant input variables, though not as highly significant as Chl-a. Because the growth and reproduction of phytoplankton rely heavily on the availability of various nutrients, the significance of nutrients can be easily understood. Moreover, the DO level is also closely associated with algal growth dynamics under subtropical coastal water conditions with mariculture activities. This is because DO is required for the respiration of organisms and also for some chemical reactions. Xu *et al.* (2004) observed that PO_4, TIN and DO have an increasing trend from the early 1970s to the late 1980s or the early 1990s, and then decline in the 1990s. As discussed in the above section, the decline can be attributed mainly to the implementation of THAP in 1988. Our training data in this case study are from 1988 to 1992, during which the trends are increasing for PO_4, TIN and DO. The significance of these variables appears to be reasonable.

It has been mentioned in the earlier section that the daily values are acquired by linear interpolation of the biweekly water quality data. It should be observed that once interpolation is employed to generate time-series from longer sampling frequency to a shorter time step, future observations are, in effect, employed to drive the predictions (Lee *et al.* 2003). One of the approaches to refrain from using this "future data" in the predictions is to forecast the algal dynamics with a lead-time at least equal to the time interval of the original observation. Thus, in this case study, a biweekly or greater lead-time forecast would be sufficient to eliminate this "interpolation effect". It is worthwhile undertaking a biweekly forecast employing the same data and identifying the significant input variables again, similar to steps performed in the one-week forecast. In this biweekly forecast, a time lag of 14–20 days is applied for each of the input variables, and Chl-a is again forecasted at time *t*. Table 7.2 lists the significant variables for the biweekly forecast. Similar to the one-week forecast, those input variables with contributions of more than 2 per cent in terms of ANN weights are shaded. From Tables 7.1 and 7.2, it can be noted that the significant input variables for biweekly forecast are almost the same as those for the one-week forecast. The only exception is Temp, which is shown to be significant in the biweekly analysis but not in the one-week forecast. Now that the biweekly predictions are free of the interpolation effect and the results are similar to those for the one-week forecast, it is justifiable to state that the significant input variables from one-week predictions are reasonable. Though the one-week forecast is to some extent driven by interpolation of data, it still contains adequate physical knowledge establishing a cause–effect relationship between the time-lagged input variables and future algal biomass.

The results of one-week-ahead predictions employing ANN are discussed in more detail in the next section. In order to reduce the computational

Table 7.2 Contribution factors on the basis of the weights of the trained ANN for biweekly predictions

Input variables	Contribution factors of the input variables (%)*							Sum
	(t−14)	(t−15)	(t−16)	(t−17)	(t−18)	(t−19)	(t−20)	
Chl-a	2.79	1.54	1.23	1.24	1.07	0.54	2.46	10.87
TIN	5.53	2.34	1.27	1.02	1.91	2.68	3.71	18.46
DO	2.42	1.96	1.86	1.39	0.63	0.80	1.19	10.26
PO₄	3.35	1.92	0.85	0.28	0.42	0.93	2.04	9.78
SD	3.50	2.38	1.38	0.20	1.06	2.27	3.66	14.45
Temp	4.04	3.09	2.39	1.78	1.18	0.94	1.36	14.79
Rain	1.89	1.55	1.23	1.08	1.38	1.26	1.57	9.96
SR	0.70	0.45	0.59	0.40	0.47	0.27	1.59	4.46
WS	1.19	0.63	1.14	0.83	1.00	1.02	1.15	6.95

Sum of contribution factors of all variables = 100

* Shaded variables have a contribution factor greater than 2%.

effort, only the variables considered as significant, namely, Chl-a, PO$_4$, TIN, DO and SD, are applied as input variables for the predictions. Seven to thirteen days of time-lagged inputs are employed for each significant input variable.

7.5.5 Results and discussion

In the following, different neural network runs, with different combinations of the significant input variables, are performed. The concentration of Chl-a is forecast with a one-week lead-time. These neural network simulations are performed with selection of input variable based on the model formulation and training details delineated in the previous section.

Results indicate that the best forecast is the case with solely time-lagged Chl-a as an input. Figure 7.4 shows the comparison of the simulated and observed Chl-a values under both training and testing conditions with solely time-lagged Chl-a as inputs. Figure 7.5 shows a blow-up of the forecast for a shorter period, i.e. May 1993 to September 1994, under the testing condition.

Table 7.3 shows the goodness-of-fit measures for the Chl-a forecast under different scenarios. It is apparent that the forecast performance deteriorates with an increase in the number of input variables, and that the best performance occurs when time-lagged Chl-a is employed as, the sole input

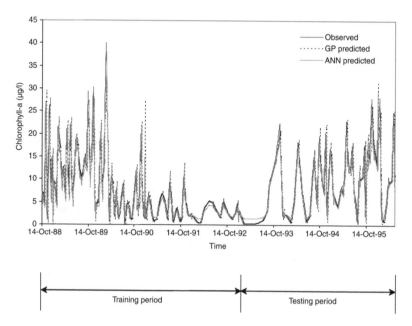

Figure 7.4 One-week forecasting of chlorophyll-a by employing the ANN model

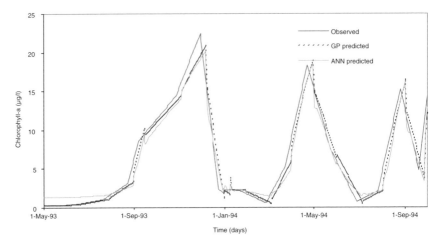

Figure 7.5 Blow-up of one-week forecasting of chlorophyll-a by employing the ANN model

Table 7.3 Performance measures for the ANN one-week prediction

Input variables*	Training		Testing	
	RMSE	CC	RMSE	CC
Chl-a, PO₄, TIN, DO, SD	1.87	0.96	4.02	0.82
Chl-a, PO₄, TIN, DO	2.16	0.94	4.49	0.86
Chl-a, PO₄, TIN	2.14	0.95	3.00	0.91
Chl-a, PO₄	2.42	0.94	2.76	0.92
Chl-a	2.55	0.93	2.24	0.95

* All input variables are at 7–13 days' time lag.

variable. This result is contrary to some previous research studies, which endeavoured to incorporate a multitude of input variables. For example, Jeong *et al.* (2003) initially selected 19 input variables, though at the end of their study, suggested that only four of them were needed to forecast biovolume of cyanobacteria with good performance; Jeong *et al.* (2001) adopted 16 input variables to forecast time-series changes of algal biomass by using a time-delayed recurrent neural network; Wei *et al.* (2001) incorporated eight environmental input variables to forecast the evolution of four dominant phytoplankton genera by using an ANN; Recknagel *et al.* (1997) selected ten input variables in a feed-forward ANN model for the forecast of algal bloom in lakes in Japan, Finland and Australia; Yabunaka *et al.* (1997) adopted ten environmental variables as ANN inputs to forecast the concentrations of five freshwater phytoplankton species. The result that solely algal biomass with a time lag is adequate for forecasting the biomass itself might cast doubt on the necessity to purchase expensive equipment such as automatic nutrient analysers for ammonia and nitrate nitrogen in coastal waters algal bloom warning detecting systems.

It is seen from the time-series plots in Figure 7.4 that the forecast can track the algal dynamics with reasonable degree of accuracy. Nevertheless, it can be observed, from a closer examination of the forecast blow-up shown in Figure 7.5, that there is a phase error of one week or so. Hence it might not be appropriate to use these biweekly data for short-term forecasts of algal blooms. It is recommended to use input data at a higher frequency in order to enhance the performance of the forecast. In fact, biweekly forecasts with the same significant input variables are also performed in this study, and the phase error is even more than its counterpart of the one-week forecast.

7.6 Prototype application II: long-term prediction of discharges

Numerical models are often used to predict both long-term flow discharges in reservoirs. Results of the forecasts can be widely employed for purposes including environmental protection, flood prevention, drought protection, reservoir control, water resource distribution, etc. They therefore have a significant impact on decision control for reservoirs and hydropower stations which might have high economic value. Conventional methods used to forecast the long-term discharges include factor analysis, hydrological analysis methods, historical evolution method, time-series analysis, multiple linear regression method, etc. The current most popular methods are time-series analysis and the multiple linear regression method. The basis of time-series analysis is the decomposition of various factors into trend and cycle. Autoregressive moving-average models as suggested by Box and Jenkins (1976) are also widely employed. The ANN model has been gaining more applications in the forecast of discharges since the 1990s. The ASCE Task Committee (2000a, 2000b) provided a comprehensive review of the application of ANNs to this field. An application of an ANN in the long-term prediction of flow is presented in the following section. Evaluation of the prediction effectiveness of the ANN model is investigated for the prototype case study in Manwan hydropower station.

7.6.1 *Scaled conjugate gradient (SCG) algorithm*

The SCG algorithm (Fitch *et al.* 1991; Hagan *et al.* 1996) is used in this case study. The procedure is as follows (Moller 1993):

1. The weight matrix w is initialized, ranging from -0.5 to 0.5:

$$\vec{d}_0 = -\vec{g}_0 \tag{7.14}$$

    where g_0 is the gradient of the error function and d_0 is an initialized searching direction.

2. At the commencement of the kth generation, the learning rate α_k is determined from linear search via the function $f\left(\vec{w}+\alpha_k\vec{d}_k\right)$, where $\vec{d}_k$ is the searching direction at the the kth generation. Adjustment of the weight matrix is made via the following equation:

$$w_{k+1}=w_k+\alpha_k\vec{d}_k \qquad (7.15)$$

3. If either the error is less than the threshold value or the predetermined training generation is reached at the $(k+1)$th generation, the training process is stopped automatically.

4. Otherwise, the new searching direction $\vec{d}_{k+1}$ is computed. If $(k+1)$ is an integer multiple of the dimension number of the weight matrix w, then

$$\vec{d}_{k+1}=-\vec{g}_{k+1} \qquad (7.16)$$

Otherwise,

$$\vec{d}_{k+1}=-\vec{g}_{k+1}+\beta_k\vec{d}_k \qquad (7.17)$$

$$\beta_k=\frac{\left(g_k g_k^T\right)}{\left(g_0 g_0^T\right)} \qquad (7.18)$$

5. Loop back to step 2.

7.6.2 *Prediction of discharges in Manwan hydropower station*

In this case study, a three-layer feed-forward backpropagation ANN model is used to forecast discharge in Manwan. There are four input neurons (Q_t, Q_{t-1}, Q_{t-2} and Q_{t-3}), four hidden neurons, and one output neuron (Q_{t+1}). Huang and Foo (2002) opined that the SCG algorithm converges faster and acquires a higher accuracy compared with other training algorithms. The daily flow measurements from 2001 to 2003 and monthly flow data from 1953 to 2003 are investigated. Preprocessing of all data is made to the raw data so that they are normalized to range from -1 to 1.

In the ANN model, daily data from 2001 are employed for training purposes whilst those from between 2002 and 2003 are used for verification. As can be seen in Figure 7.6, the minimum error is only 0.00598 after training for 361 epochs. During the verification period, the correlation coefficient between the forecast and measured values and the RMSE are 0.97 and 0.0087, respectively, as shown in Figure 7.7. It is shown that the forecast of daily flow results is excellent. In a similar fashion, monthly data from between 1953 and 1993 are employed for training purposes, whilst those from between 1994 and 2003 are for verification. It can be shown that the forecast of monthly flow results is satisfactory.

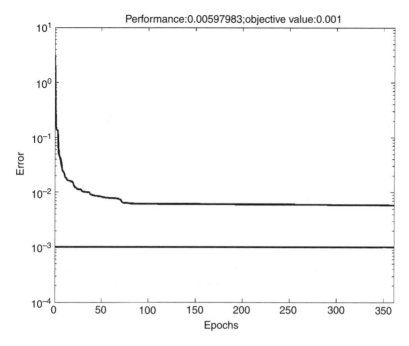

Figure 7.6 Training results of daily flow data at Manwan Hydropower by the ANN
   model

7.6.3 Results and discussion

It is generally recognized that there are many available algorithms in training ANNs, and that each of them has its own advantages and drawbacks.
Hence, it is worthwhile comparing the performance of the gradient descent,
LM and SCG algorithms in this case study. In order to have the same basis
of comparison, all three algorithms undergo the training process under the
same conditions. Table 7.4 details the performance comparison of various
algorithms for monthly flow prediction in Manwan. Results show that the
gradient descent algorithm has the slowest convergence, the smallest correlation coefficient and the largest RMSE. Amongst the three algorithms, the
SCG algorithm is the most accurate and the LM algorithm converges in the
fastest manner.

Amongst various available time-series prediction models, the auto-
regression time-series model is very often used. It has been employed
conventionally in flow prediction owing to the simplicity of both the model
structure and the data requirements (Wang 2000). It is therefore employed
as the benchmarking tool in order to gauge the performance of this ANN
model for flow prediction in Manwan. The same training and verification
sets are used for both models so as to have the same basis for comparison.

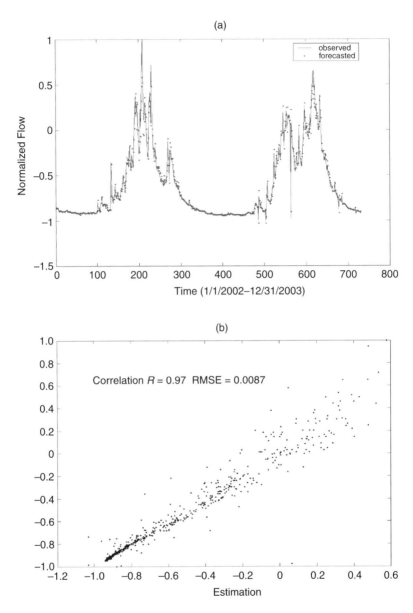

Figure 7.7 (a) and (b) Verification results of daily discharge at Manwan Hydropower by ANN model

Table 7.4 Performance comparison of different training algorithms for monthly flow forecasting at Manwan Hydropower

Training algorithm	Correlation coefficient	Normalized RMSE	Number of sampling points
Gradient descent	0.799	0.057	1000
LM	0.878	0.036	71
SCG	0.890	0.03	415

Table 7.5 Performance comparison between the ANN model and the time-series model for the forecasting of monthly discharge at Manwan Hydropower

Correlation coefficient		Normalized RMSE	
ANN model	Time-series model	ANN model	Time-series model
0.89	0.84	0.03	0.108

In this case study, results indicate that the ANN exhibits distinct advantages over conventional time-series models in terms of accuracy performance. Table 7.5 presents a performance comparison of the ANN model and the time-series model for monthly flow forecast in Manwan. Whilst the correlation coefficient of the ANN model, 0.89, is larger than that of the time-series model, 0.84, the RMSE of the ANN model, 0.03, is much smaller than its counterpart of the time-series model, 0.108.

7.7 Conclusions

The first application case study presents the analysis of algal dynamics data from a coastal monitoring station using ANN. The interpretation of ANN weights is employed to identify the most significant input variables, which appear to be consistent with ecological reasoning. Results indicate that chlorophyll-a alone with a time lag is sufficient as input for forecasting itself. The phenomenon suggests an auto-regressive nature of the algal dynamics in this semi-enclosed coastal zone. Results show that although the use of biweekly data can mimic the long-term trends of algal biomass in a reasonable manner, it might not be appropriate to use these data for the short-term forecast of algal blooms. It is recommended to use input data at a higher frequency in order to enhance the performance of the forecast.

Secondly, an ANN model is employed to forecast long-term flow discharges in Manwan on the basis of historical records. The daily flow measurements from 2001 to 2003 and monthly flow data from 1953 to 2003 are investigated. The results indicate that the ANN model can give good prediction performance. The correlation coefficients between the forecast and field values are 0.97 and 0.89 for daily and monthly discharges, respectively. Different training algorithms are tried, with performance compared. It is found that the SCG algorithm is the most accurate amongst them. Moreover, results indicate that ANN exhibits distinct advantages over a conventional time-series model in terms of accuracy performance.

8   Fuzzy inference systems

8.1 Introduction

Fuzzy logic and fuzzy set theory were introduced by Zadeh (1965). There has been a rapid growth in the number and variety of their applications during the past few decades. The applications range from consumer products such as cameras, washing machines and microwave ovens to industrial process control, decision-support systems, medical instrumentation, portfolio selection, etc. A recent trend is the use of fuzzy logic in combination with neurocomputing and genetic algorithms. In fact, fuzzy logic, neurocomputing and genetic algorithms are all principal constituents of soft computing. Amongst various hybrid combinations of methodologies in soft computing, an appropriate combination of fuzzy logic and neurocomputing should be highlighted. The resulting adaptive-network-based fuzzy inference system (ANFIS) is an effective system that operates through a parallel and fault-tolerant architecture on both linguistic descriptions of the parameters and the numeric values. In this chapter, the characteristics of fuzzy logic, fuzzy inference systems and ANFIS are delineated. A real application of ANFIS is also demonstrated, for the prediction of long-term flow discharges in Manwan based on historical records.

8.2 Fuzzy logic

Fuzzy logic is very useful in simulating imprecise and complex systems (Zadeh and Kacprzyk 1992). Under fuzzy set theory, mapping is made from elements of a fuzzy set to a universe of membership values using a function-theoretic form belonging to the close interval between zero and one. A key process in applying fuzzy methods is the determination of the membership function of a variable. This process is analogous to the estimation of probability in stochastic models. Membership functions in fuzzy set theory, having the capability of simulating the preferences of the decision-maker, can be acquired according to real statistical surveys. It is quite simple to model based on fuzzy logic since it only operates on an "if–then" principle, i.e. "if" is a vector of fuzzy explanatory variables or premises with membership functions and "then" is a consequence. Both the "if" and the "then" clause are in the form of a fuzzy set.

An optimization problem having a vague objective or constraints can be solved as a fuzzy optimization problem. Fuzzy set theory can furnish an alternative approach to addressing those problems without well-defined objectives and constraints or without precise information. Although fuzzy logic has been employed in a multitude of applications, it is often employed as a refinement to conventional optimization techniques in which fuzzy constraints are used to replace the conventional crisp objective and some or all of the constraints (Cheng and Chau 2001; Cheng *et al.* 2002). Silvert (1997) applied fuzzy set theory concepts to ecological impact classifications. Chang *et al.* (2001) employed the fuzzy synthetic evaluation approach to identify the quality of river water. Chen and Mynett (2003) employed data mining techniques and heuristic knowledge in modelling the fuzzy logic of eutrophication in Taihu Lake. Liou *et al.* (2003) applied a two-stage fuzzy set theory to evaluate river quality in Taiwan. Marsili-Libelli (2004) delineated the design of a bloom predictor based on the daily fluctuations of simple parameters for water quality such as dissolved oxygen, oxidation–reduction potential, pH, and temperature.

A comparison between the simulated and measured results of flow or water quality is a typical example of the application of fuzzy logic. Improvements in modelling results rely heavily on the technology of pattern recognition. The normalized root-mean-square error (NRMSE) between key field data and the modelling results is computed to evaluate the performance of the model and its pertinent model parameters. The NRMSE might include different cases, namely, a time-series of data at a single location within the model domain, instantaneous measurements at many locations, or a combination of both of the above cases. In this case, the expression of NRMSE is as follows:

$$\text{NRMSE} = \frac{\sum\limits_{i=1}^{N}\sum\limits_{t=1}^{n}(T_{i,t} - O_{i,t})^2}{\sum\limits_{i=1}^{N}\sum\limits_{t=1}^{n}(T_{i,t} - \overline{T})^2} \tag{8.1}$$

where N is the number of spatial data locations for comparison, n is the number of time intervals in a time-series of data for comparison, $T_{i,t}$ is the target values of the ith spatial location and tth time step, $O_{i,t}$ is the computed value of the ith spatial location and tth time step, and $\overline{T}$ is the average target value.

Figure 8.1 shows the membership functions for NRMSE, in which the fuzzy logic of literal classification is represented by four categories: very small, small, large, and very large. Another application of fuzzy inference can be shown in the representation of rule sets within the knowledge

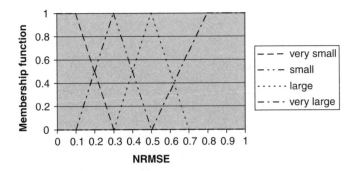

Figure 8.1 Membership functions for the normalized root-mean-square error with four categories

base of a knowledge-based system. In this case, a more human-like fuzzy format, instead of a crisp threshold format, can be employed.

The study of fuzzy logic and fuzzy set theory was pioneered by Zadeh (1965) in the 1960s. It was commonly used in modelling ambiguity and uncertainty in decision-making. The key idea for fuzzy logic is the allowance for something to be partial truth, instead of having to be just either "true" or "false" in a crisp manner. Similarly, in fuzzy set theory, it is possible to have partial belonging to a set, usually termed a fuzzy set. The degree of "belongingness" to a set or category is represented numerically by a membership function. Its range lies between 0 and 1.0 and its type or shape might be triangular-shaped, trapezoidal-shaped, bell-shaped, and so on. As a demonstration, a triangular membership function, which is the simplest and most popularly employed, can be expressed as follows (Shrestha *et al.* 1996):

$$\mu_T(x) = \frac{x - t_1}{t_2 - t_1} I_{t_1,t_2}(x) + \frac{t_3 - x}{t_3 - t_2} I_{t_2,t_3}(x) \tag{8.2}$$

where $\mu_T(x)$ = grade of membership of x in T with $t_1 \leq t_2 \leq t_3$, $I_{(...)}(x)$ = an indicator function, with value equal to non-zero when x is within the interval indicated and zero otherwise, and $\mu_T(x)$ is greater than zero within the interval (t_1, t_3).

In general, fuzzy logic programming can be employed in two different ways. The first is to endeavour to simulate the behaviour of a human expert whilst the second is to map a set of outputs to a set of inputs in a fuzzy inference method (Russell and Campbell 1996). In order to simulate the thinking of a human expert, input variables are usually specified by category, such as "low", "high"; and fuzzy rules are developed according to the expert's knowledge and experience. However, if no expertise is available, the number of membership functions or the number of categories of input variables assigned to each input variable can be selected empirically.

8.3 Fuzzy inference systems

A fuzzy inference system (FIS) executes a non-linear mapping between its input space and the output space. This mapping is attained by a number of fuzzy if–then rules, each of which delineates the local behaviour of the mapping. In fuzzy modelling, the parameters of the if–then rules, usually termed "antecedents" or "premises", specify a fuzzy region of the input space whilst the output parameters, usually named "consequents", define the corresponding output. The process of fuzzy inference involves fuzzy logic operators, membership functions, if–then rules, etc. As shown in Figure 8.2, the basic structure of a FIS consists of three conceptual components: a rule base, a database, and a reasoning mechanism. The rule base comprises a selection of fuzzy rules. The database defines the membership functions (MF) employed in the fuzzy rules. The reasoning mechanism executes the inference procedure upon the rules in order to determine an output. Three types of fuzzy inference systems are often employed nowadays: Mamdani-type (Mamdani and Assilian 1975), Sugeno-type (Takagi and Sugeno 1985; Sugeno and Kang 1988) and Tsukamoto-type (Tsukamoto 1979). The key difference of these three types of inference systems is basically in the manner of determination of outputs.

The FIS is similar to an ANN in terms of the conceptual set-up. Their objectives are both to identify the transformation of a set of inputs to the corresponding set of outputs through "training" or "learning". On the other hand, an ANN tends to behave more like a "black box" operation whilst a fuzzy logic system is more transparent. This is because an expert's knowledge and experience can be incorporated into the inference process when it is required. Jang (1993) delineated different types of fuzzy inference systems

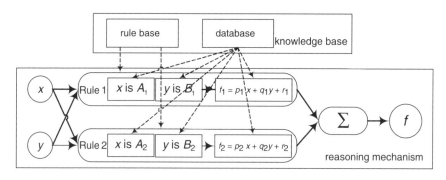

Figure 8.2 Basic structure of fuzzy inference system

and categorized them into three types on the basis of the types of fuzzy reasoning and fuzzy if–then rules employed. Takagi and Sugeno's fuzzy if–then rules are used in the application case study. The output of each rule is the summation of a linear combination of input variable and a constant term whilst the final output is the weighted averaged of each rule's output. Figure 8.3 shows the fuzzy reasoning for two input variables.

A fuzzy rule base has to be established after fuzzy reasoning has been defined. Usually, the fuzzy rule base can be constructed according to expert knowledge or measured data. In cases of lack of expertise, the simplest approach can be used, i.e. the rule base is established by combining all categories of variables. An exemplary case with three input variables and a single output variable is illustrated in the following paragraphs.

It is assumed that the three input variables x, y and z are each divided into three categories and that equally spaced triangular membership functions are adopted. Literal description can be given to indicate the attributes of the categories, such as "low", "medium" and "high." As a general rule, the number of rules in the fuzzy rule base is c^n, where c denotes the number of categories per variable and n represents the total number of variables. Certainly there is no limitation on the number of categories. The optimal

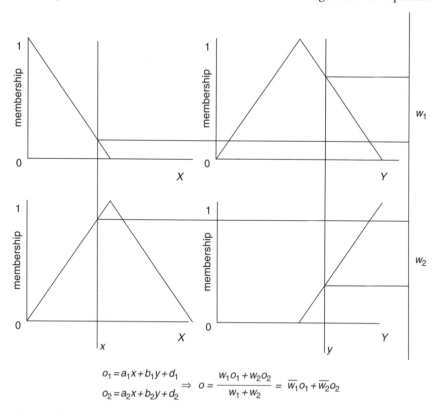

$$o_1 = a_1 x + b_1 y + d_1 \\ o_2 = a_2 x + b_2 y + d_2 \Rightarrow o = \frac{w_1 o_1 + w_2 o_2}{w_1 + w_2} = \overline{w}_1 o_1 + \overline{w}_2 o_2$$

Figure 8.3 Fuzzy reasoning for two input variables

number of categories can be determined via a trial and error procedure, after having compared the simulation result with different numbers of categories. The rule base takes the form of an output $o_{i,j,k}$ for any combination of category i, of input variable x, category j, of input y, and category k, of input variable z, amounting to a total of 27 rules ($= 3^3$) in this case. The following is part of the rule set:

> If x is low, y is low, and z is low then the output $o_{1,1,1} = a_1x + b_1y + c_1z + d_1$;
> If x is low, y is low, and z is medium then the output $o_{1,1,2} = a_2x + b_2y + c_2z + d_2$;
> If x is low, y is low, and z is high then the output $o_{1,1,3} = a_3x + b_3y + c_3z + d_3$;
>
> $\vdots$
>
> If x is high, y is high, and z is high then the output $o_{3,3,3} = a_{27}x + b_{27}y + c_{27}z + d_{27}$;

where $a_{(...)}, b_{(...)}, c_{(...)}$, and $d_{(...)}$ are parameters of fuzzy output functions. After the training process of the adaptive networks, these parameters can be determined.

Once a rule is triggered, memberships for x, y, and z are computed. The result of the application of a specific T-norm operation, such as multiplication, max, min, and square, will furnish the weight $w_{i,j,k}$ to be assigned to the corresponding output $o_{i,j,k}$. In the application case study in this chapter, multiplication operation is adopted. At the end, the outputs from all rules that are triggered are coupled to furnish a single weighted average output as follows:

$$o = \frac{\sum w_{i,j,k} \cdot o_{i,j,k}}{\sum w_{i,j,k}} \tag{8.3}$$

where i, j and k are index of categories of x, y and z, respectively. Following the above procedure, the values of the output o can be computed for all possible combination of values of variables x, y and z. Over the past two decades, FIS has been successfully applied in fields including decision analysis, expert systems, automatic control, data classification, and computer vision.

8.4 Adaptive-network-based fuzzy inference system (ANFIS)

In order to render the FIS model a prediction model, these parameters, including t_1, t_2 and t_3 of each triangular membership function and a_i, b_i, c_i and d_i of the consequence part of each rule, have to be determined by some learning laws. If a neural network is employed to train the parameters of the FIS, it is known as a neural FIS. Jang (1993) integrated an FIS with an adaptive network, termed an ANFIS. The underlying principle of the ANFIS was that a FIS was mapped into a neural network structure and the parameters were optimized using a hybrid learning algorithm in accordance with the least squares approach and backpropagation learning. As a matter of fact, a multitude of learning algorithms, such as simply the backpropagation algorithm, could be employed instead. However, as demonstrated by Jang (1993), this hybrid learning algorithm had the apparent advantage of quick convergence.

Since then, the ANFIS has been successfully applied to different domains, including prediction of workpiece surface roughness (Lo 2003), pesticide prediction in ground water (Sahooa *et al.* 2005), validation in financial time-series (Koulouriotis *et al.* 2005), etc. In particular, a neuro-fuzzy system for modelling time-series on flow discharges was presented by Nayak *et al.* (2004). Ponnambalam *et al.* (2002) employed an ANFIS for minimization of the variance of reservoir systems operations benefits, and attained a satisfactory result. This is generally recognized that a fuzzy system based on hybrid algorithms can enhance the intelligence of systems working in uncertain, imprecise, and noisy environments. This is because it possesses the attributes of both systems, namely, learning abilities, optimization abilities, and connectionist structures for neural networks, and also human-like "if–then" rule reasoning, and the readiness to integrate expert knowledge for fuzzy systems.

8.4.1   ANFIS architecture

In the Sugeno model (or Takagi–Sugeno model) proposed by Takagi and Sugeno (1985), a typical rule in a Sugeno fuzzy model has the form

If x is A and y is B, then $z = f(x,y)$

where A and B denote fuzzy sets of antecedent, and $z = f(x,y)$ represents the precise function. Often $z = f(x,y)$ are polynomials of input variables x and y. The function $z = f(x,y)$ is a first-order polynomial of the input variables in the commonly used first-order Sugeno model. The output level z is set to be a constant for a zero-order Sugeno model. For a first-order Sugeno fuzzy model having two inputs x and y and one output z, a typical rule set with two fuzzy if–then rules can be expressed as:

Rule 1: If x is A_1 and y is B_1, then $f_1 = p_1 x + q_1 y + r_1$

Rule 2: If x is A_2 and y is B_2, then $f_2 = p_2 x + q_2 y + r_2$

Figure 8.4 shows the fuzzy reasoning mechanism for this Sugeno model employed to determine an output function (f) from a known input vector $[x, y]$. The Sugeno FIS, which is computationally efficient, can work together

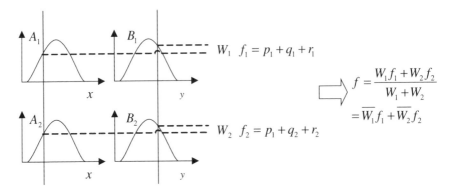

Figure 8.4 Fuzzy reasoning mechanism for Sugeno model used in an ANFIS model

with other linear techniques, optimization and adaptive techniques. When an FIS is developed employing the framework of adaptive neural networks, it is known as an ANFIS.

In a typical ANFIS configuration, the parameters defining the shape of the membership functions and the consequent parameters for each rule are determined by the backpropagation learning algorithm and the least-squares method, respectively. In the following section, the neuro-fuzzy network is a five-layer feed-forward network that employs neural network learning algorithms integrated with fuzzy reasoning in order to map an input space to an output space. Figure 8.5 shows the ANFIS architecture with details as described below.

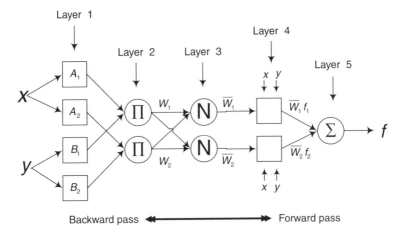

Figure 8.5 ANFIS architecture with five layers

Layer 1: input nodes

Each node in this layer generates the membership grade of an input variable. The output of the node, $O_{1,i}$, is determined by

$$O_{1,i} = \mu_{A_i}(x), \; i = 1, 2 \tag{8.4}$$

or

$$O_{1,i} = \mu_{B_{i-2}}(y), \; i = 3, 4 \tag{8.5}$$

where x (or y) is the input to the node; A_i (or B_{i-2}) is a fuzzy set associated with this node. This fuzzy set is dependent on the shape of the MFs in the node, which can be any appropriate functions that are continuous and piecewise differentiable such as Gaussian, generalized bell shaped, trapezoidal shaped or triangular shaped functions. If a generalized bell function is adopted as the MF, the output $O_{1,i}$ becomes

$$\mu_A(x) = \frac{1}{1 + \left| \dfrac{x - c_i}{a_i} \right|^{2b}} \tag{8.6}$$

where $\{a_i, b_i, C_i\}$ is the parameter set that determines the actual shapes of the MF which is subjected to the limiting range between 0 and 1; and $\{a_i, b_i, c_i\}$ are termed premise parameters.

Layer 2: rule nodes

The node in this layer is multiplied to the incoming signals, represented as $\prod$. The output $O_{2,i}$, representing the firing strength of a rule, is defined to be

$$O_{2,i} = w_{A_i}(x) \mu_{B_i}(y), \; i = 1, 2 \tag{8.7}$$

Thus, the outputs $O_{2,i}$ of this layer represent the products of the corresponding degrees from layer 1.

Layer 3: average nodes

The node of this layer, termed N, denotes the normalized firing strengths as follows:

$$O_{3,i} = \overline{w} = \frac{w_i}{w_1 + w_2}, \; i = 1, 2 \tag{8.8}$$

Layer 4: consequent nodes

Node i in this layer assesses the contribution of the ith rule towards the model output, with equation below:

$$O_{4,i} = \overline{w}_i f = \overline{w}_i (p_i + q_i + r_i) \tag{8.9}$$

where $\overline{w}_i$ is the output of layer 3 and $\{p_i, q_i, r_i\}$ is the consequent parameter set.

Layer 5: output nodes

The only node in this layer that determines the overall output of the ANFIS is represented by the following:

$$\text{Overall output} = O_{5,1} = \sum \overline{w}_i f_i = \frac{\sum_i w_i f_i}{\sum_i w_i} \tag{8.10}$$

8.4.2 Hybrid learning algorithm

The ANFIS architecture comprises two parameter sets for optimization: the premise parameters $\{a_i, b_i, c_i\}$, which define the shape of the MFs, and the consequent parameters $\{p_i, q_i, r_i\}$, which delineate the overall output of the system. It can be observed from the ANFIS architecture in Figure 8.5 that when the values of the premise parameters are known, the overall output can be written as a linear combination of the consequent parameters. The output f in Figure 8.5 can be expressed symbolically as

$$f = \overline{w}_1 f_1 + \overline{w}_2 f_2$$
$$= (\overline{w}_1 x) p_1 + (\overline{w}_1 y) q_1 + (\overline{w}_1) r_1 + (\overline{w}_2 x) p_2 + (\overline{w}_2 y) q_2 + (\overline{w}_2) r_2 \tag{8.11}$$

which is a linear function in terms of the consequent parameters $p_1, q_1, r_1, p_2, q_2, r_2$. As such, a hybrid learning algorithm, which integrates the backpropagation gradient descent and the least squares estimate method, has the capability of outperforming the original backpropagation algorithm (Rumelhart *et al.* 1986). The consequent parameters are first updated employing the least squares algorithm. The antecedent parameters are then updated by backpropagating the errors. In particular, during the forward pass of the hybrid learning algorithm, node outputs proceed until layer 4 and the consequent parameters are determined by the least squares method. During the backward pass, the error signals propagate backwards and the premise parameters are then updated by gradient descent algorithm. Table 8.1 gives a summary of the activities in both forward and backward

Table 8.1 Summary of the activities in forward and backward passes for ANFIS

	Forward pass	*Backward pass*
Premise parameters	Fixed	Gradient descent
Consequent parameters	Least-squares estimate	Fixed
Signals	Node outputs	Error signals

passes. Jang and Sun (1995) furnish more details about the hybrid learning algorithm.

8.5 Advantages and disadvantages of fuzzy inference systems

It is generally recognized that, in coastal engineering problems, many indicators are often in conflict with each other, significant observations may be lacking, and potentially valuable information may be non-quantitative in nature. One of the advantages of the fuzzy inference method is that it is capable of representing real-life hydrodynamic and water quality problems, which are often difficult to deal with by standard mathematical and statistical approaches. Nevertheless, fuzzy logic also has its drawbacks. By itself, it cannot help much with user-friendly interactions between the users and the system. Moreover, the choice of many parameters, including the number of categories, shape of the membership function, and method of combining partial memberships, can greatly affect the results. For accurate representations of the simulation, it is necessary to have the appropriate choice of all pertinent parameters as well as to undergo rigorous validation. This might not be an easy task, particularly for novice users.

8.6 Applications and case studies

In this case application, the ANFIS is used to predict flood propagation in a channel reach. The study site is at Manwan Hydropower in the Lancangjiang River, which is a large river in Asia. It originates from the Qinghai-Tibet Plateau, penetrates Yunnan from the north-west to the south, flows through several countries, namely, Laos, the Union of Myanmar, Thailand, Cambodia and Vietnam, and ultimately exits into the South China Sea. The total length of the Lancangjiang River is 4,500 miles or so, covering a catchment area of approximately 744,000 square miles. The Manwan Hydropower is located at the middle reaches of the Lancangjiang River between the counties of Yunxian and Jingdong. The drainage area at the Manwan dam site is 114,500 square miles. The total length of river reach upstream of Manwan amounts to 1,579 miles, with a mean elevation of 4,000 metres. At the dam site, the average yearly runoff is 1,230 cubic metres, most of which is contributed by rainfall whilst about 10 per cent comes from snow melt. Moreover, most of the annual rainfall, amounting to 70 per cent, is precipitated during the wet season, which is between June and September each year.

Figure 8.6 shows the monthly flow data at Manwan Reservoir from January 1953 to December 2003, which are investigated in this study. The data set from 1953 to 1998 and that from 1999 to 2003 are employed for training and validation, respectively. Following the suggestion by Masters (1993), the data sets of river flow are preprocessed and normalized to within the range of zero to one prior to the simulation process.

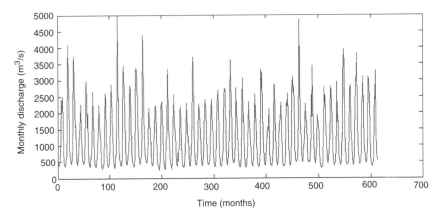

Figure 8.6 Monthly discharge data of the Manwan Reservoir from 1953 to 2003

8.6.1 *Model development and testing*

There are no universal rules for developing an ANFIS, although a general framework can be figured out following previous successful applications in various fields. It is apparent that a key objective of an ANFIS is to generalize a relationship of the form of

$$Y = f(X^n) \tag{8.12}$$

where X^n is an n-dimensional input vector comprising variables $x_1, \ldots, x_i, \ldots, x_n$, and Y is the output variable. In modelling of flow data series, values of x_i are usually flow values at different time lags, and the value of Y is in general the flow at the imminent period. Yet, the optimal number as well as which antecedent flow values showed be incorporated in the vector X^n so as to attain the best performance is unknown a priori. Therefore, in this study, a trial and error procedure is adopted. Initially, an ANFIS model is set up with a single antecedent flow in the input vector. Then, a new ANFIS model is developed one at each time, with the input vector being adjusted by successively adding flow at another time lag. Altogether, six ANFIS models are developed as follows:

$$\text{Model } n \quad Q_t = f(Q_{t-1} \ Q_{t-n}) \ n = 1, \ldots, 6$$

where Q_t corresponds to the river flow discharge at time t.

The model performance is evaluated in terms of the following performance indices:

1. The coefficient of correlation (CORR) given by

$$
\text{CORR} = \frac{\frac{1}{n}\sum_{i=1}^{n}\left(Q_o(i) - \overline{Q_o}\right)\left(Q_f(i) - \overline{Q_f}\right)}{\sqrt{\frac{1}{n}\sum_{i=1}^{n}\left(Q_o(i) - \overline{Q_o}\right)^2} \times \sqrt{\frac{1}{n}\sum_{i=1}^{n}\left(Q_f(i) - \overline{Q_f}\right)^2}}
\tag{8.13}
$$

where $Q_o(i)$ and $Q_f(i)$ are the observed and predicted discharges, respectively, $\overline{Q_o}$, $\overline{Q_f}$ denote the mean observed and predicted discharges, respectively, and n is the total number of data points.

2. The root mean square error (RMSE) given by

$$
\text{RMSE} = \sqrt{\frac{1}{n}\sum_{i=1}^{n}(Q_f(i) - Q_o(i))^2}
\tag{8.14}
$$

8.6.2 *Results and discussion*

Table 8.2 shows the resulting performance indices of ANFIS, namely, CORR and RMSE, employing the Gaussian membership function and the trapezoidal membership function, respectively, for all models. The membership function of every input parameter within the architecture can be divided into two categories: small and large. The results illustrate that Model 3, comprising antecedent flows at three different time lags in input, acquires the maximum CORR and minimum RMSE during validation, no matter whether the membership function is of Gaussian or trapezoidal shape. As such, Model 3 with three antecedent input data is chosen as the best-fit model for delineating the flow of the Manwan Hydropower. Moreover, the effect of the choice of membership function on the model performance is also investigated. In this case, six different membership functions are attempted using Model 3: the triangular membership function (TRIMF), the trapezoidal membership function (TRAPMF), the generalized bell membership function (GBELLMF), the Gaussian membership function (GAUSSMF), the Gaussian combination membership function (GAUSS2MF), the spline-based membership function (PIMF) and the sigmoidal membership function (DSGMF). Table 8.3 presents the result performance of Model 3 coupling with different membership functions. It can be observed that the TRAPMF performs the best, with the maximum CORR and minimum RMSE during validation, whilst the GBELLMF attains the worst result.

Table 8.2 Performance indices for Models 1–6 by using two types of membership functions

Model	Gaussian membership function				Trapezoidal membership function			
	Training		Validation		Training		Validation	
	RMSE	CORR	RMSE	CORR	RMSE	CORR	RMSE	CORR
1	0.11843	0.78539	0.13043	0.77773	0.11889	0.78348	0.12958	0.78156
2	0.090325	0.88157	0.10475	0.86359	0.09186	0.87722	0.10694	0.85762
3	0.075927	0.91793	0.099208	0.87957	0.075795	0.91823	0.097094	0.88877
4	0.06605	0.93861	0.13718	0.78263	0.067406	0.93597	0.10266	0.87995
5	0.061604	0.9469	0.14105	0.78515	0.065892	0.939	0.16199	0.72977
6	0.058825	0.9518	0.41629	0.38461	0.060644	0.94868	0.27504	0.58358

Table 8.3 Performance indices for Model 3 by using different membership functions

Membership function	Training		Validation	
	RMSE	CORR	RMSE	CORR
TRIMF	0.079641	0.9093	0.097281	0.88339
TRAPMF	0.075795	0.91823	0.097094	0.88877
GBELLMF	0.075036	0.91993	0.10304	0.86983
GAUSSMF	0.075927	0.91793	0.099208	0.87957
GAUSS2MF	0.074961	0.9201	0.098256	0.88327
PIMF	0.075463	0.91898	0.98573	0.88652
DSGMF	0.07424	0.92169	0.99168	0.87999

8.6.3 Result comparison with an ANN model

It is noted that, during the past two decades, the ANN model has been commonly applied in flow prediction. One of the advantages of the ANN approach over conventional mechanistic models is that it is not necessary to represent information about the complex nature of the underlying process explicitly in mathematical form. Therefore an ANN model is established employing the same input parameters to the ANFIS Model 3 in order to act as a benchmarking model for comparison purposes. In the ANN model, the scaled conjugate gradient algorithm (Fitch *et al.* 1991; Moller 1993) is used for training, and the optimized number of hidden neurons is determined by trial and error procedure. The optimized ANN architecture comprises three hidden neurons. The same training and verification sets are used for both models so as to have the same basis of comparison. Figures 8.7 and 8.8

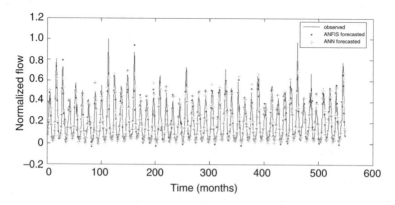

Figure 8.7 Result comparison of ANFIS and ANN flow forecasting models during training period (1953–1998)

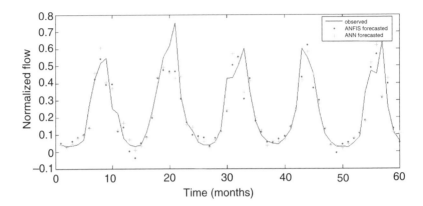

Figure 8.8 Result comparison of ANFIS and ANN flow forecasting models during validation period (1999–2003)

Table 8.4 Performance indices of ANN and ANFIS models

	Training		Validation	
	RMSE	CORR	RMSE	CORR
ANFIS	0.075795	0.91823	0.097094	0.88877
ANN	0.080755	0.90662	0.099927	0.87766

show the performances of both ANN and ANFIS models during training period and validation periods, respectively. Table 8.4 presents the performance indices of both ANN and ANFIS models during the training and validation periods. It can be observed that, for flow forecasts in Manwan, ANFIS exhibits some advantages over the ANN model in terms of model performance. The correlation coefficient of the ANFIS model during validation period, having a value of 0.88877, is higher than that of the ANN model, which is 0.87766. In terms of the RMSE, the value for ANFIS model is 0.097094, which is better than its counterpart of the ANN model of 0.099927.

8.7   Conclusions

In the case study in Manwan Hydropower, an ANFIS model is employed to forecast long-term flow discharges on the basis of available historical records. The monthly data are split into two parts: data between 1953 and 1998 are employed for training whilst data between 1999 and 2003 are employed for validation. Results illustrate that the ANFIS model is capable of furnishing good forecast performance. The RMSE are 0.075795 and 0.097094 for training and validation, respectively whilst the correlation coefficients between the forecast and the measured values are 0.91823 and 0.88877 for training and validation, respectively. Different membership functions and different numbers of input variables at successively increasing numbers of time lags for ANFIS have been tried, which indicate that a combination of the TRAPMF and three antecedent flows in input produces the best performance in the monthly forecast of discharges in Manwan Hydropower. Moreover, a benchmarking comparison is made with an appropriate ANN model, which indicates that the ANFIS model is capable of furnishing an even better performance. This illustrates its distinct capability and advantages in simulating flow discharge times series with non-linear attributes.

9   Evolutionary algorithms

9.1   Introduction

Evolutionary algorithms employ computational models of natural evolutionary processes in developing problem-solving systems (Goldberg 1989). This form of search evolves throughout generations by enhancing the attributes of potential solutions and simulating the natural population of biological entities. In this chapter, several types of evolutionary algorithms, including genetic algorithms (GA), genetic programming (GP), and particle swarm optimization (PSO), are delineated. Three real applications of evolutionary algorithms are also demonstrated. The first application case study presents the use of GP for modelling and prediction of algal blooms in Tolo Harbour, Hong Kong. The second application is for flood forecasting at a prototype channel reach of the Yangtze River in China by employing a GA-based artificial neural network (ANN) in comparison with several benchmarking models. The third application is the use of a PSO training algorithm for ANNs in stage prediction of Shing Mun River.

9.2   Genetic algorithms (GA)

Holland (1975) pioneered the GA as an optimization method to minimize or maximize an objective function. It is a powerful global search algorithm based on the concepts of natural genetics and the mechanisms of biologically inspired operations (Holland 1992). An extensive description of GA can be found in Goldberg (1989). GA in essence applies the concept of the artificial survival of the fittest integrated with a structured information exchange employing randomized genetic operators simulated from nature in order to formulate an efficient search mechanism. GA works on the collective learning process within a population of individuals, each of which denotes a search point in the space of potential solutions. There are several differences of GA from conventional optimization algorithms, namely, working on a coding of parameter sets, population processing, probabilistic operators, and separation of domain knowledge from search.

Moreover, GA is not constrained by assumptions about continuity or existence of derivatives.

GA endeavours to exploit efficiently useful information subsumed in a population of solutions with better performance. It accomplishes this objective by using a multitude of operations to generate a novel and enhanced population of strings from an old population. The iterative process to generate and test a population of strings simulates a natural population of biological entities where successive generations of entities are conceived, born, and brought up until they are ready to reproduce. GA searches from a population of strings and overcomes many peaks in parallel at the same time, thus lowering the chance of locating local optima. GA entails that alternative solutions are coded as strings. These strings may consist of concatenation of some substrings, each of which denotes a design variable. Different coding schemes have been employed successfully in solving different types of problems. In a population of strings, individuals and characters are known as chromosomes and artificial genes, respectively.

The first step in a GA is the random generation of an initial population. The fitness value of each individual, which is a measure of optimality of the objective function, is then evaluated. It is this criterion that gauges the effect of the population's evolution employing different operators in order to generate new and hopefully better solutions. Various genetic operators that have been identified and employed in GAs comprise reproduction, crossover, deletion, dominance, intra-chromosomal duplication, inversion, migration, mutation, selection, segregation, sharing, translocation, etc. Amongst them, reproduction, crossover and mutation are in common use. Figure 9.1 shows a typical flow chart detailing how a typical GA generates the solution.

Reproduction is tailored to employ fitness to guide the evolution of chromosomes. The reproduction operator works on the principle of survival of the fittest in the population. Amongst many strings, an old string is copied into the new population on the basis of its fitness value. Hence, under this operator, strings with better objective function values, denoting more highly fit, get more offspring in the mating pool. In fact, there are many methods to implement the reproduction operator and any one of them that biases selection toward fitness can be applicable. Crossover is the process by which chromosomes chosen from a source population are mixed to generate offspring to become potential members of a successor population under the hope that quality offspring might be produced by quality parents. The crossover operator results in the rearrangement of individual genetic information from the mating pool and the generation of new solutions to the problem. This operator is applied to each offspring in the population with a preset crossover probability. A variety of crossover operators, such as uniform, single point, two points and arithmetic crossover, are available. A mutation operator is employed to maintain the diversity in the population and to keep away from local minima by preventing the individuals in

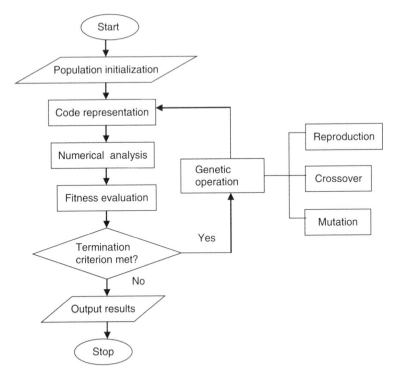

Figure 9.1 Flow chart of a typical GA

a population from becoming too closely related. This operator is applied to each offspring in the population with a preset mutation probability.

Since the early studies on the subject, different applications has been made and GAs have illustrated their capacity to yield good solutions even in highly complex and multiple-parameter domains (Chau and Albermani 2003; Cheng *et al.* 2002). GAs were shown to be able to reveal the existence of patterns, regularities and relationships that drive a certain phenomenon, such as algal abundance. Mulligan and Brown (1998) employed GAs to calibrate water quality models. Bobbin and Recknagel (2001a) employed a GA to build inducing explanatory rules for the forecasting of algal blooms. Ng and Perera (2003) used GAs to calibrate a river water quality model. Cho *et al.* (2004) employed GAs to optimize regional wastewater treatment in a river water quality management model.

In coastal modelling, as a typical example, a GA was employed to determine an appropriate combination of parameter values in a flow and water quality mechanistic model (Chau 2002). It should be noted that many model parameters might not be directly obtained from field measurements, and that the inappropriate use of their values might generate large errors or even lead to numerical instability. The percentage errors of peak value, peak time and total volume of coastal constituents were key performance measures for model predictions. The calibration of parameters relied on observed data on tidal as well as water quality constituents gleaned over a five-year period in the Pearl River. Another two-year record was employed to verify these parameters. A sensitivity analysis on crossover probability, mutation probability, population size, and maximum number of generations was also undertaken in order to determine the most fitting algorithm parameters. Results illustrate that the application of a GA was capable of simulating the important characteristics of the coastal process and that the calibration of models was efficient and robust. Readers are referred to Chau (2002) for more details.

9.3 Genetic programming (GP)

The basic search strategy behind genetic programming (GP), as a type of evolutionary algorithm, is similar to that of a GA, which imitates biological evolution as described in the last section. The key difference of GP from a conventional GA is that it operates on parse trees rather than bit strings (Koza 1992). In GP, a parse tree is set up from a terminal set, which denotes the variables in the problem, and a function set. For illustration purposes, a simple case is considered here in which the terminal set comprises a single variable x and some constants, and the function set comprises the operators for multiplication, division, addition and subtraction. Then, the space of available parse trees comprises all polynomials of any form over x and the constants. Figure 9.2 shows one of the parse trees for the model representing the equation of $y = -0.2x + 0.3$.

GP proceeds similarly to GA by initially producing a population of random parse trees, computing their fitness as a measure of how well they fit the given problem, and then selecting the better parse trees for reproduction and rearrangement of chromosomes to constitute a new population of offspring. Iteration of this process of selection and reproduction is made

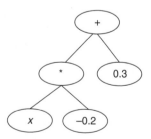

Figure 9.2 Example of GP parse tree denoting the equation of $-0.2x + 0.3$

until some termination criterion is reached. The rearrangement of chromosomes is made by a crossover operator, representing random swapping of sub-trees of the parse trees between selected individuals at a predetermined crossover probability. Readers are referred to Babovic and Abbott (1997), Babovic and Keijzer (2000) and Bobbin and Recknagel (2001b) for more comprehensive descriptions of GP.

9.4   Particle swarm optimization (PSO)

The particle swarm optimization (PSO) algorithm was originally developed as a tool for modelling social behaviour. It is found to have the ability to optimize complex numerical functions (Kennedy and Eberhart 1995; Kennedy 1997). PSO lies somewhere between evolutionary programming and GA, and is another type of evolutionary technique under the domain of computational intelligence (Clerc and Kennedy 2002). It is an optimization paradigm that simulates the ability of human societies to process knowledge with roots in two key component methodologies: artificial life such as bird flocking, fish schooling and swarming; and evolutionary computation (Clerc and Kennedy 2002).

The underlying principle of the PSO algorithm is based on the assumption that potential solutions will be flown through hyperspace with acceleration towards more optimum solutions. Similar to GA, it is a populated search method in acquiring optimized solutions for non-linear functions. However, it achieves this goal by resembling the movement of organisms in a bird flock or fish school, instead of using genetic operators as in GA. Particles or individuals are terms used to represent candidate solutions to the problem. In PSO, the evolution of generations of a population of these particles is accomplished solely by cooperation and competition within the group of particles themselves. In other words, in any time step, each particle endeavours to modify its flying pattern on the basis of the previous flying experiences of both itself and its companions. In the flying process, each particle keeps track of its coordinates in hyperspace associated with its previous best fitness solution, and also of the entire swarm representing the overall best value obtained thus far by any other individual within the population.

In the PSO algorithm, it is convenient to adopt vectors to represent particles, as in most optimization problems. The population is evolving, taking consideration of the quality attributes of the previous best individual values as well as the previous best group values. Moreover, the allocation of responses between the individual and group values is such that a diversity of response is ensured.

Owing to its advantage in a capability to locate the global optimum and fast convergence, it is adopted to train the multilayer perceptrons addressing matrices' learning problems. A three-layered perceptron is selected for this application case to be trained by a PSO. In the following sections,

$W^{[1]}$ and $W^{[2]}$ denote the connection weight matrix linking the input layer and the hidden layer, and that linking the hidden layer and the output layer, respectively. The ith particle in the PSO is represented by

$$W_i = \{W_i^{[1]}, W_i^{[2]}\} \tag{9.1}$$

The location denoting the previous best fitness value of any individual is kept tracked and expressed by

$$P_i = \{P_i^{[1]}, P_i^{[2]}\} \tag{9.2}$$

It is supposed that the index of the best particle among all the particles in the current population is denoted by the symbol b. Then, it is apparent that the best matrix is expressed by

$$P_b = \{P_b^{[1]}, P_b^{[2]}\} \tag{9.3}$$

The velocity of individual i is represented by

$$V_i = \{V_i^{[1]}, V_i^{[2]}\} \tag{9.4}$$

Now it is assumed that m and n denote the index of matrix row and column, respectively. Thus, the new values of the individuals become

$$
\begin{aligned}
V_i''^{[j]}(m,n) = V_i^{[j]}(m,n) \\
+ \frac{\left\{ r\alpha \left[P_i^{[j]}(m,n) - W_i^{[j]}(m,n) \right] + s\beta \left[P_b^{[j]}(m,n) - W_i^{[j]}(m,n) \right] \right\}}{t}
\end{aligned} \tag{9.5}
$$

and

$$W_i''^{[j]} = W_i^{[j]} + V_i^{[j]} t \tag{9.6}$$

where $j = 1, 2$; $m = 1, \ldots, M_j$; $n = 1, \ldots, N_j$; M_j and N_j are the row and column sizes of the matrices W, P, and V; r and s are positive constants; α and β are random numbers ranging between 0 and 1; t is the incremental time step between observations and can often be converted to unity; and V'' and W'' represent the new values of velocity and weight matrix, respectively. Equation (9.5) is used to determine the new velocity of the particle on the basis of its previous velocity and the distances of its current position from the best experiences of its own and also of the entire group. Equation (9.5) can be explained in the context of social behaviour. The equation comprises two distinct components: the cognition component, represented by the second term on the right-hand side of the equation, denoting the private thinking of the particle itself; and the social part, represented by the third element on the right-hand side, denoting the collaboration among the particles as a group. With this new velocity known, equation (9.6) computes its new position.

The equation for the fitness or objective function of the ith individual is formulated in term of an output mean squared error of the neural networks as below:

$$f(W_i) = \frac{1}{S} \sum_{k=1}^{S} \left[\sum_{l=1}^{O} \{t_{kl} - p_{kl}(W_i)\}^2 \right] \qquad (9.7)$$

where f is the fitness value, t_{kl} is the target output; p_{kl} is the predicted output based on W_i; S is the number of training set samples; and O is the number of output neurons.

9.5 Advantages and disadvantages of evolutionary algorithms

A principal advantage of GA is its ability to search for the global optimum solution to a complex problem. Nevertheless, GA is an algorithmic process so that it cannot help much to furnish too many user-friendly interactions with the users. The accuracy and performance of a prediction depend very much on the appropriate selection of various GA parameters, including crossover probability, mutation probability, population size, and maximum number of generations. In order to accomplish a reliable prediction, it is necessary to choose these parameters carefully, possibly via a trial and error procedure. Moreover, similar to many other soft computing techniques, a GA cannot extrapolate beyond the range of the training data. The major drawback of GA is that it may not necessarily result in the best possible solution because of the limitation on local searching capability.

The major advantage of GP for the modelling process is its capability of generating models that establish an intelligible structure, namely, a formula or equation. Hence GP might be particularly appropriate to deal with instances which are "data rich, theory poor", in comparison to other techniques. This is because GP can self-adapt, through the genetic loop, a population of function trees so as to ultimately formulate an "optimal" and physically interpretable model.

The key advantages of PSO are the relatively simple and computationally efficient coding and its adaptability and tracking to the change of the previous best group fitness value. The stochastic PSO algorithm was proven to be capable of locating the global optimum with high probability and high convergence rate (Clerc and Kennedy 2002).

9.6 Prototype application I: algal bloom prediction by GP

The first prototype application is for real-time algal bloom prediction at Tolo Harbour in the north-eastern coastal waters of Hong Kong by GP. The nature of the data employed and details of the modelling are shown in the following sections.

9.6.1 Description of the study site

Tolo Harbour is a semi-enclosed bay in the north-eastern coastal waters of Hong Kong and is connected to the open sea via Mirs Bay. It is generally recognized that the water quality gradually deteriorates from the better flushed outer "Channel Subzone" towards the more enclosed and densely populated inner "Harbour Subzone".

In Tolo Harbour, the nutrient enrichment exhibited by the eutrophication phenomenon resulting from municipal and livestock waste discharges has long been a major environmental concern and threat over the past few decades. Two major treatment plants at Shatin and Taipo are continuously releasing organic loads. There are also non-point sources from runoff and direct rainfall, as well as waste from mariculture. The eutrophication effect has generated frequent algal blooms and red tides, which occurred particularly in the weakly flushed tidal inlets inshore. As a consequence, occasional massive fish kills are caused owing to severe dissolved oxygen depletion or toxic red tides. Numerous studies have been made, which concluded that the state of health of the ecosystem of Tolo Harbour had been progressively deteriorating since the early 1970s (Morton 1988; Xu *et al.* 2004). During this period, serious stresses to the marine coastal ecosystem were caused by the nutrient enrichment in the harbour due to urbanization, industrialization, livestock rearing, etc. Frequent occurrences of red tides and associated fish kills were recorded in the late 1980s, which was possibly the worst period. Morton (1988) considered that the Tolo Harbour was "Hong Kong's First Marine Disaster" and that the inner harbour was effectively dead. At that time, since Tolo Harbour had entered a critical stage, the Hong Kong Government decided to develop and implement an integrated action plan, Tolo Harbour Action Plan (THAP). Significant effectiveness in the reduction of pollutant loading and in the enhancement of the water quality was recorded through the implementation of THAP in 1988.

There have been a number of field and process-based modelling research studies on eutrophication and dissolved oxygen dynamics of this harbour during the past few decades (e.g., Chan and Hodgkiss 1987; Chau and Jin 1998; Lee and Arega 1999, Chau 2004a; Xu *et al.* 2004).

The development and implementation of a GP model for real-time algal bloom prediction at Tolo Harbour is presented in this study. The data used in this study are mainly the monthly/biweekly water quality data collected by the Hong Kong Government's Environmental Protection Department as part of its routine water quality monitoring programme. The data from the most weakly flushed monitoring station, TM3, are chosen so as to separate the ecological process from the hydrodynamic effects as far as possible. All data of ecological variables are collected and presented as depth-averaged.

In order to obtain the daily values, linear interpolation of the biweekly observed data is performed. Moreover, some meteorological data including wind speed, solar radiation and rainfall recorded by the Hong Kong Observatory are employed. No interpolation is thus required on these sets of data since they are recorded in a daily manner. The data are split into two portions: data between 1988 and 1992 for training, and data between 1993 and 1996 for testing the models. Lee and Arega (1999) and Lee *et al.* (2003) furnish more detailed descriptions of the water quality data in this study site.

9.6.2   *Criterion of model performance*

The objective function used for the evolution of the GP models is the minimization of the root mean square error (RMSE) of the prediction over the training period. The evaluation of the performance of the predictions for GP is made by two goodness-of-fit measures, the root mean square error (RMSE) and correlation coefficient (CC). Visual comparison can be made on time-series plots.

$$\text{RMSE} = \sqrt{\frac{\sum_{i=1}^{p}[(X_m)_i - (X_s)_i]^2}{p}} \tag{9.8}$$

$$\text{CC} = \frac{\sum_{i=1}^{p}\left[(X_m)_i - \left(\overline{X}_m\right)_i\right]\left[(X_s)_i - \left(\overline{X}_s\right)_i\right]}{\sqrt{\sum_{i=1}^{p}\left[(X_m)_i - \left(\overline{X}_m\right)_i\right]^2 \left[(X_s)_i - \left(\overline{X}_s\right)_i\right]^2}} \tag{9.9}$$

where subscripts m and s = the measured and simulated chlorophyll-a levels, respectively; p = total number of data pairs considered; $\overline{X}_m$ and $\overline{X}_s$ = mean value of the measured and simulated data, respectively. RMSE furnishes a quantitative indication of the model error in units of the variable, with the attribute that larger errors draw greater attention than smaller ones. The coefficient of correlation between the measured and simulated data can be considered a qualitative evaluation of the model performance.

9.6.3   *Model inputs and output*

In the GP models, there are nine input variables and one output variable. Nine input variables are chosen to be the most influential factors on the algal dynamics of Tolo Harbour:

1. chlorophyll-a, Chl-a (μg/L);
2. total inorganic nitrogen, TIN (mg/L);

3. dissolved oxygen, DO (mg/L);
4. phosphorus, PO$_4$ (mg/L);
5. secchi-disc depth, SD (m);
6. water temperature, Temp (°C);
7. daily rainfall, Rain (mm);
8. daily solar radiation, SR (MJ/m^2); and
9. daily average wind speed, WS (m/s).

This is determined after taking careful consideration of previous computer modelling and field observation studies in the weakly flushed embayment in Tolo Harbour (Chau *et al.* 1996; Lee and Arega, 1999; Lee *at al.* 2003). The sole model output is chlorophyll-a, which serves as a good indicator of the algal biomass. The model is applied to attain one-week prediction of algal blooms. This time period is selected after having taken consideration of the ecological process at Tolo Harbour and practical constraints of data collection. A time lag of 7–13 days is introduced for each of the input variables. The time lag has to commence from seven for a targeted one-week lead-time of the forecasting.

9.6.4 *Significant input variables*

The selection of the appropriate model input variables is very significant for GP, similar to other machine learning (ML) techniques such as ANN. As shown in the last section, the selection of input variables is in general according to a priori knowledge of causal variables and physical/ecological insight of the modeller into the problem domain. Moreover, the use of lagged input variables results in better predictions in a complex and dynamical system.

In GP models, evolution of the GP equations is performed in order to determine an empirical relationship between the nine selected input variables with time lag of 7–13 days and the chlorophyll-a concentration at time *t*. For each of the nine input variables, there are seven time-lagged variables, rendering a total of 63 ($= 9 \times 7$) input variables. In order to reduce the computational effort, there is a necessity to determine, from these 63 variables, the significant input variables.

In this study, an endeavour is made to employ the evolutionary search capabilities of GP to choose the significant input variables. Table 9.1 lists the GPKernel parameters employed for all the GP runs. In the table, the parameters "maximum initial tree size" and "maximum tree size" denote the maximum size of the tree of the initial population and of the population in subsequent generations, respectively. Their corresponding values are constrained to 45 and 20, respectively. This constraint is imposed because GP tends to evolve uncontrollably large trees if there is no limitation on the tree size. In fact, a maximum tree size of 20 furnishes simple expressions that are intelligible. It is noted that, when "maximum tree size" is limited to

Table 9.1 Values of key parameters employed
            in GP runs

Parameter	Value
Maximum initial tree size	45
Maximum tree size	20
Crossover rate	1
Mutation rate	0.05
Population size	500
Elitism used	Yes

20, there will only be four to eight variables in the evolved equation. Therefore the evolutionary process will select only about four to eight significant
variables from the total of 63 input variables.

Preprocessing of the data is required so as to refrain from acquiring a
functional relationship comprising dimensionally non-homogeneous terms
within the evolved GP model. All the variables are therefore normalized or
non-dimensionalized initially, which can be performed by simply dividing
all the variables by their respective maximum values.

GPKernel, which was developed by DHI Water & Environment, is available at http://www.d2k.dk, is the software employed in this study. GPKernel
is a command line based tool for finding functions on data. All computations were carried out on a Pentium PC with 1021 MB RAM for each
adopted function and variable set, and GPKernel is executed for 30 CPU
minutes in order to acquire the optimal solution.

Table 9.2 lists the four different function sets employed for the GP runs.
For each function set, 20 GP equations are evolved by employing different
initial seeds. Hence a total number of 80 GP equations were evolved for
the one-week forecast. It can be observed that small and simple function
sets are employed. The main reason is that GP is very creative at captur-
ing simple functions and generating whatever it requires by combination.
During the evolution process, GP often ignores the more complicated functions in favour of the simple ones (Banzhaf *et al.* 1998). Moreover, a simple
function set results in the evolution of simple GP models, which are more

Table 9.2 Function sets employed
            for the GP runs

Function set
$+, -, {}^\times, /$
$+, -, {}^\times, /, e^x$
$+, -, {}^\times, /, x^2$
$+, -, {}^\times, /, x^y$

Table 9.3 Number of input variable selections in 80 GP runs for one-week forecst

Input variables	Number of terms of time-lagged input variables[*]							Total terms
	$(t-7)$	$(t-8)$	$(t-9)$	$(t-10)$	$(t-11)$	$(t-12)$	$(t-13)$	
Chl-a	229	53	65	58	51	46	23	525
TIN	14	7	10	1	3	2	5	42
DO	18	14	5	5	10	6	14	72
PO$_4$	38	10	9	4	2	4	1	68
SD	13	17	4	11	2	1	2	50
Temp	4	3	2	1	5	7	5	27
Rain	0	0	0	0	0	1	1	2
SR	0	0	1	0	0	2	1	4
WS	0	0	0	0	0	0	0	0
					Total number of terms in 80 GP models=			790

[*] Shaded variables contribute to more than 2% of the 790 terms in the 80 GP models

intelligible. It is believed that GP has the capability of choosing the input variables which are beneficial to the model and thus the GP evolved equations would contain the most significant ones out of the 63 input variables. In other words, the number of times the variable is selected in the evolved equations should give a good indication of its significance.

Table 9.3 lists the total number of times each of the 63 input variables is selected in the 80 evolved equations. In the table, the significant variables, which are defined to be those with number of terms more than two per cent of the total number of terms in the 80 GP equations, are shaded. In this case study, the total number of terms in the 80 equations is 790 and variables that contribute more than two per cent of 790 are those with 16 or more terms. It can be observed from this analysis that Chl-a at $t-7$ is the most significant constituent in predicting the one-week-ahead algal biomass. Other variables are PO$_4$, DO, TIN and SD, having a contribution factor larger than 2 per cent.

It is apparent from Table 9.3 that, apart from Chl-a at $t-7$ being the most significant in predicting itself, all Chl-a values during the previous one to two weeks are also significant. This indicates an auto-regressive nature or "persistence" for the algal dynamics, which is common for geophysical time-series. Such an auto-regressive nature is frequently exhibited by these types of time-series owing to their inherent inertia or carryover process in the physical system. Numerical modelling can contribute to the comprehension of the physical system by revealing factors about the process that establish persistence into the series. In this case study, the long flushing time or residence time in the semi-enclosed coastal waters might be the key factor causing the auto-regressive nature of chlorophyll dynamics as revealed by the ML techniques. This was also proposed in a recent ANN study of algal dynamics in Tolo Harbour (Lee *et al.* 2003). It should be pointed out that, in the inner and outer Channel Subzones, the tidal currents are very small, with the average current velocities being 0.04 m/s and 0.08 m/s, respectively (EPD 1999). The weak tidal flushing in the harbour is therefore due to its landlocked nature. Lee and Arega (1999) suggested that the flushing times in the inner Harbour Subzone were in the order of one month, and hence the retention time was quite long.

It can be observed from the results that the nutrients, i.e. PO_4 and TIN, together with DO and SD to a lesser extent, are also significant input variables, though not as highly significant as Chl-a. Because the growth and reproduction of phytoplankton rely heavily on the availability of various nutrients, the significance of nutrients can be easily understood. Moreover, the DO level is also closely associated with algal growth dynamics under subtropical coastal waters conditions with mariculture activities. This is because DO is required for the respiration of organisms and also for some chemical reactions. Xu *et al.* (2004) observed that PO_4, TIN and DO have an increasing trend from the early 1970s to the later 1980s or the early 1990s, and then decline in the 1990s. As discussed in the above section, the decline can be attributed mainly to the implementation of THAP in 1988. Our training data in this case study are from 1988 to 1992, during which the trends are increasing for PO_4, TIN and DO. Thus it is reasonable that these variables are significant.

It should be noted that the daily values are acquired by linear interpolation of the biweekly water quality data. It should be observed that once interpolation is employed to generate time-series from longer sampling frequency to a shorter time step, future observations are, in effect, employed to drive the predictions (Lee *et al.* 2003). One of the approaches to refrain from using these "future data" in the predictions is to forecast the algal dynamics with a lead-time at least equal to the time interval of the original observation. Thus, in this case study, a biweekly or more lead-time forecast would be sufficient to free from this "interpolation effect". It is worthwhile undertaking a biweekly forecast employing the same data and identifying the significant input variables again, similar to steps performed in the one-week forecast. In this biweekly forecast, a time lag of 14–20 days is applied for each of the input variables and Chl-a is again forecasted at time t.

Table 9.4 lists the significant variables for the biweekly forecast by GP. Similar to the one-week forecast, those input variables with contribution more than two per cent in terms of number of terms in GP equations are shaded. From Tables 9.3 and 9.4, it can be noted that the significant input variables for biweekly forecast are almost the same as those for the one-week forecast. The only exception is Temp, which is shown to be significant in the biweekly analysis but not in the one-week forecast. Now that the biweekly predictions are free of the interpolation effect, the results are similar to those for the one-week forecast. It is thus justifiable to state that the significant input variables from one-week predictions are reasonable.

Table 9.4 Number of input variable selections in 80 GP runs for biweekly forecast

Input variables	Number of terms of time-lagged input variables*							Total terms
	$(t-14)$	$(t-15)$	$(t-16)$	$(t-17)$	$(t-18)$	$(t-19)$	$(t-20)$	
Chl-a	140	18	71	111	13	9	6	368
TIN	10	5	4	1	0	4	10	34
DO	31	10	11	2	3	3	11	71
PO$_4$	24	10	2	1	3	9	15	64
SD	35	9	3	5	2	0	7	61
Temp	14	8	2	3	2	1	9	39
Rain	0	0	0	1	0	0	2	3
SR	0	1	3	11	10	11	13	49
WS	0	0	0	1	0	1	0	2
					Total number of terms in 80 GP models=			691

* Shaded variables contribute to more than 2% of the 691 terms in the 80 GP models

Though the one-week forecast is to some extent driven by interpolation of data, adequate physical knowledge establishing a cause–effect relationship between the time-lagged input variables and future algal biomass is still built-in.

In the next section, the results of one-week ahead predictions employing GP are discussed in more details. In order to reduce the computational effort, only the variables considered as significant, namely, Chl-a, PO$_4$, TIN, DO and SD, are applied as input variables for the predictions. For each significant input variable, 7–13 days of time-lagged inputs are included.

9.6.5 Results and discussion

During the evolution of GP models, the concentration of Chl-a is forecast with a one-week lead-time by applying the significant input variables delineated in the previous section. Five different GP runs are performed with each of the four function sets as shown in Table 9.2. Almost the same controlling parameters as those employed for the input variable selection shown in Table 9.1 are used for the GP runs. The only exception is the parameter "maximum tree size", which is increased to 45 in order to attain a lower RMSE.

Results show that the best GP model, as indicated by the one with minimum RMSE, was evolved with the function set comprising the basic math operators (+, −, ×, /). This reinforces our belief of the understanding that GP entails simple function sets to generate models with the best forecasting capability and also gives confirmation to our decision to adopt simple function sets as shown in Table 9.1. Figure 9.3 shows a comparison of the simulated and observed Chl-a values under both training and testing conditions with solely time-lagged Chl-a as inputs, which is the

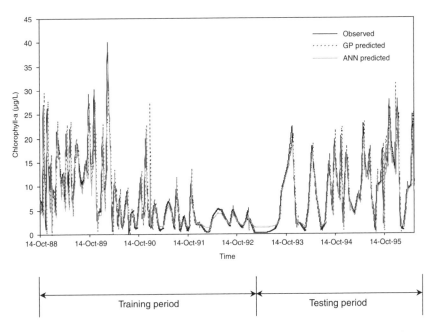

Figure 9.3 One-week forecasting of chlorophyll-a by employing GP model

best GP model. Figure 9.4 shows a blow-up of the forecast for a shorter period.

Table 9.5 shows the performance measures for the Chl-a forecast under different scenarios for GP predictions. It is apparent that the forecast performance deteriorates with increase in the number of input variables, and that the best performance occurs when time-lagged Chl-a is employed as sole input variable. This result is in contrast to some previous research studies, which endeavoured to incorporate a multitude of input variables. For example, Jeong *et al.* (2003) initially selected 19 input variables, though at the end of their study suggested that only four of them were needed to forecast biovolume of cyanobacteria with good performance; Jeong *et al.* (2001) adopted 16 input variables to forecast time-series changes of algal biomass by using a time-delayed recurrent neural network; Wei *et al.* (2001) incorporated eight environmental input variables to forecast the evolution of four dominant phytoplankton genera by using an ANN; Recknagel *et al.* (1997) selected ten input variables in a feed-forward ANN model for the forecast of algal bloom in lakes in Japan, Finland and Australia; Yabunaka *et al.* (1997) adopted ten environmental variables as ANN inputs to forecast the concentrations of five freshwater phytoplankton species. The result that solely algal biomass with a time lag is adequate for forecasting the biomass itself might cast doubt on the need to purchase other expensive equipment

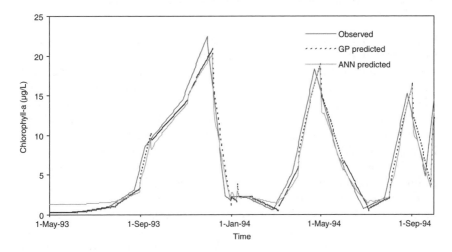

Figure 9.4 Blow-up of one-week forecasting of chlorophyll-a by employing GP model

such as automatic nutrient analysers for ammonia and nitrate nitrogen as part of algal bloom warning detecting systems in coastal waters.

Results from the time-series plots in Figure 9.3 indicate that the forecast can track the algal dynamics with a reasonable degree of accuracy. Nevertheless, it can be observed, from a closer examination of the forecast blow-up as shown in Figure 9.4, that there is a phase error of one week or so. Hence it might not be appropriate to use these biweekly data for short-term forecasting of algal blooms. It is recommended to use input data at a higher frequency in order to enhance the performance of the forecast. In fact, biweekly forecasts with the same significant input variables are also performed in this study. The phase error is even more than its counterpart of the one-week forecast and hence is not shown.

Table 9.5 Performance measures for the one-week forecast by GP

Input variables*	Training		Testing	
	RMSE	CC	RMSE	CC
Chl-a, PO_4, DO, SD	2.67	0.92	2.54	0.93
Chl-a, PO_4, DO	2.55	0.92	2.50	0.93
Chl-a, PO_4	2.37	0.93	2.32	0.94
Chl-a	2.55	0.93	1.99	0.95

* All input variables are of 7–13 days time lag

9.7 Prototype application II: flood forecasting in river by ANN-GA

It is generally recognized that a conventional backward propagation neural network with a gradient descent learning algorithm has the drawback of slow convergence, thus entailing very long computational time. On the other hand, GA is known to have the capability to search for the global optimum solution to a complex problem, but it may not necessarily result in the best possible solution owing to a shortage of the capability of local searching, which is just the advantage of ANN. Hence, in this section, a hybrid learning algorithm (ANN-GA) is developed for flood prediction. A GA is used to optimize initial parameters including weights and biases of an ANN whilst training is continued by the ANN itself. This section presents the application of a hybrid algorithm, namely, a genetic algorithm-based artificial neural network (ANN-GA), for water stage/flooding prediction in a reach of the Yangtze River in China. Several benchmark models, namely, a linear regression (LR) model, a conventional ANN model, and a conventional GA model, are employed to gauge the performance of this ANN-GA model.

9.7.1 Algorithm of ANN-GA flood forecasting model

GA, which applies biological principles to computational algorithms to attain the optimum solutions, is a robust method for searching for the optimum solution in a complex and dynamic problem. Though it may not be able to result in the best feasible solution in all cases, it can often accomplish the required accuracy (Goldberg and Kuo 1987). For comparison of its performance against a linear model, a linear model with GAs for optimizing parameters is expressed below:

$$x_{t+1} = ax_t + bx_{t-1} + cx_{t-2} - d \tag{9.10}$$

where a, b, c and d are parameters.

The objective of this problem is to determine the optimal parameters so that cumulative errors between simulated and measured data are minimal. Thus, the fitness function can be written as follows:

$$f(a,b,c,d) = \sum_{i=1}^{p} \sum |(X_m)_i - (X_s)_i| \tag{9.11}$$

In this case study, the ranking selection method (Baker 1985) is adopted and the probability prob(rank) is expressed as follows:

$$\text{prob}(\text{rank}) = q(1-q)^{\text{rank}-1} \tag{9.12}$$

where q is a user-defined parameter, and rank is the position of an individual ranked in either ascending or descending order. The objective of the ranking selection is to furnish a higher chance for a good chromosome to be chosen for the next generation. After the computation of prob(rank), roulette wheel selection, which is based on cumulative prob(rank), is employed here (Goldberg and Deb 1989). Moreover, a two-point crossover and a simple mutation operator are adopted.

A genetic-algorithm-based artificial neural network (ANN-GA) model is developed and implemented here. It is possible that a hybrid integration of ANN and GA algorithms may furnish better performance by taking advantage of the characteristics of both algorithms. This ANN-GA model might be able to speed up the convergence of an ANN model and also enhance the local searching capability of a GA model at the same time. In this algorithm, a GA is used to optimize initial parameters including weights and biases of the ANN whilst training is continued by the ANN itself. The objective of the GA sub-model is to determine optimal parameters so as to accomplish minimal cumulative errors between the simulated and actual data. The fitness function of the GA sub-model employed for initializing weights and biases is written as below:

$$\min J(W,\theta) = \sum_{i=1}^{p} |Y_i - f(X_i, W, \theta)| \qquad (9.13)$$

where W is the weight, θ is the bias or threshold value, i is the data sequence, p is the total number of training data pairs, X_i is the ith input data, Y_i is the ith measured data, and $f(X_i, W, \theta)$ denotes simulated output. Figure 9.5 shows the overall flow chart of the ANN-GA model, where p_c is the crossover probability, p_m is the mutation probability, $G_{\max}$ is the maximum number of generation, and $N_{\max}$ is the population size.

9.7.2 The study site and data

The ANN-GA model is applied to a river reach in the middle portion of the Yangtze River, as shown in Figure 9.6. The Yangtze is the largest river in China and passes through the capital of Hubei province, Wuhan. The Yangtze River is characterized by intrinsically unsteady but roughly seasonal flow behaviour. Generally speaking, the peak flow and the dry weather flow happen during the summer and winter months, respectively. Hence water year is roughly divided into two seasons: a wet season and a dry season, between June and October and between November and the next May, respectively. As a typical example to illustrate the drastic change in water stage, its value at Luo-Shan station is at 17.3 m during the dry season, but becomes 31.0 m during the wet season. The mean water elevations at this station during the dry and wet seasons are 20.8 m and 27.1 m, respectively.

The objective of this study is to predict the water levels of the downstream station, Han-Kou, based on the known water levels of the upstream

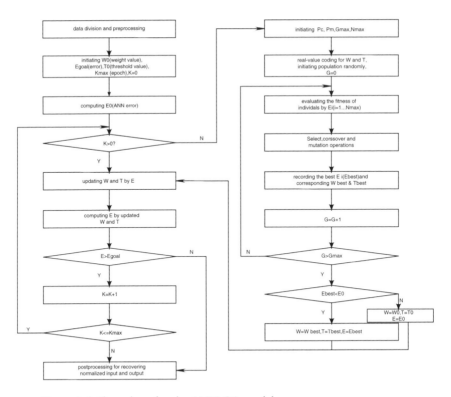

Figure 9.5 Flow chart for the ANN-GA model

station, Luo-Shan, at different lead times. The lateral inflow is neglected since its value is small in comparison with the discharge of the main stream. By using the Muskingum method, coupled with on-site measured data, the travel time of flood between Luo-Shan and Han-Kou is determined to be about 24 hours. In other words, the phase difference between the flood wave at Han-Kou and its counterparts at Luo-Shan is one day or so. It is believed that water elevations during the previous few days at Lou-Shan will exert some effects to the water stage at Han-Kou. Hence it is feasible to determine the correlation function between a time-series having D points of time spacing Δ apart, $x\,[t-(D-1)\,\Delta]\,,\,\cdots\,,x\,(t-\Delta)\,,x\,(t)$ and a predicted value $x(t+p)$ at a prescribed time in the future. According to the data availability and the phase lag found between the two locations, the following values are adopted for these parameters: $p=1$ day and $\Delta=1$ day. Since it is expected that the choice of D will have a significant impact on the results, a trial and error method is undertaken in order to find the optimal value of D. In this case study, values from 1 to 4 are attempted. Daily averaged water stages of the Luo-Shan and Han-Kou stations in 1984, 1985, 1986 and 1987 are employed in the model.

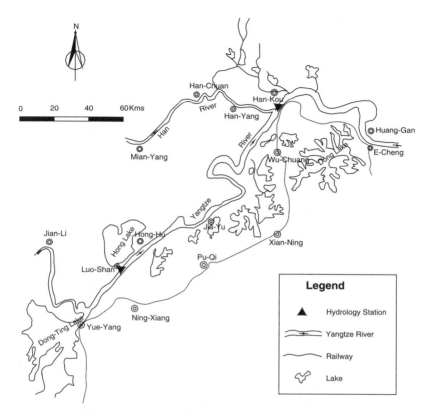

Figure 9.6 Map showing studied reach in the Yangtze River

The over-fitting problem is often considered a big challenge in pattern recognition. This problem occurs when the output endeavours to fit the training data too well. Both the underlying mapping and the noise are mimicked. Thus, when the model is applied to a new set of data with different noise, the fitting is not good. Smith (1993) proposed a multitude of techniques to overcome this problem, such as limiting the number of hidden nodes, adopting smaller weights, and limiting the number of training epochs. Shahin *et al.* (2002) recommended splitting the raw data into three sets, which is followed in this case study. Preprocessing is made to the data, which are randomly split into three independent sets, namely, training, testing, and validation sets, with proportions of 50 per cent, 25 per cent and 25 per cent, respectively. Hence, from the entire data record, 1,456 input-output data pairs are set up with the following format:

$$\{x\,[t-(D-1)],\ldots,x\,(t-2),x\,(t-1),x\,(t),x\,(t+1)\} \tag{9.14}$$

Table 9.6 Statistical parameters for training, testing, and validation sets at Luo-Shan station

Data sets	Statistical parameters		
	Mean	Standard deviation	Range
Training set	23.44	3.71	17.35–31.04
Testing set	23.44	3.71	17.39–30.96
Validation set	23.44	3.71	17.37–30.93

which denotes the correlation among water stages at Luo-Shan during the previous few days and the water elevation at Han-Kou in the ensuing day. In order to ensure that the model is not required to extrapolate beyond the range of the training data, investigations have been made to analyse the training, testing and validation sets. The statistical parameters, including the mean, standard deviation, minimum, maximum and range for the training, testing, and validation sets respectively are shown in Table 9.6. It can be seen that the criterion is satisfied.

As in Application I in Section 9.6, two popular performance measures, namely, the RMSE and CC, are used to gauge the goodness-of-fit of the forecast resulting from training, testing and validation. RMSE furnishes a quantitative indication of the model error in units of the variable, with the attribute that larger errors draw greater attention than smaller ones. The coefficient of correlation between the measured and simulated data can be considered a qualitative evaluation of the model performance.

9.7.3   Results and discussion

At first, a conventional GA model is used with a floating-point coding. Hence, each chromosome comprises four variables, a, b, c and d. The ranges of each variable are preset with reference to coefficients of the LR model: the range of a, b, c is between -2.0 and 2.0, and the range of d is between -10.0 and 10.0. After a trial and error process, the following values of the parameters are adopted: the size of population $\text{pop}_{\text{size}} = 300$, the crossover probability $p_c = 0.9$, mutation probability $p_m - 0.1$, and $q = 0.08$. Following the initiation step, the genetic operations are applied. In each generation, $p_c \times \text{pop}_{\text{size}}$ and $p_m \times \text{pop}_{\text{size}}$ chromosomes are randomly selected for crossover and mutation operation, respectively. The reproduction operation is according to the cumulative prop(rank), as expressed in equation (9.12). Since the objective is to minimize the cumulative errors, a smaller value of the fitness function denotes a higher rank. The optimal values for a, b, c and d are determined to be 1.620, -1.005, 0.395 and -5.073, respectively. Table 9.7

Table 9.7 Comparison of different values for *D* in GA model

D	Training set		Validation set	
	RMSE	CC	RMSE	CC
4	0.235	0.9958	0.245	0.9957
3	0.240	0.9959	0.238	0.9960
2	0.2417	0.9958	0.243	0.9958
1	0.2423	0.9958	0.244	0.9958

shows the performance of the GA model. It can be observed that the GA model accomplishes the best performance when *D* equals 3, which is similar to the LR model. Because of the high linearization of this problem, the advantage of GAs over LR might not be revealed. In fact, a key advantage of a GA is its robustness in searching for the optimum solution for a complex and non-linear problem. Nevertheless, this case study indicates its characteristic of acquisition of a comparatively near-to-global optimal solution, but not a guarantee of the most optimal result. This is possibly because GA involves so many random operations, including selection, initialization, and crossover and mutation operations. Hence, although it is able to acquire a comparatively near-to-global optimal solution, it is not easily once to search for the optimal solution randomly.

In the ANN-GA model, three inputs and one output are adopted so as to provide comparison on the same basis. A trial and error procedure is used to determine the optimal architecture of the ANN-GA models with different number of hidden nodes ranging from 1 to 7, which is 3-3-1. Table 9.8 shows the performances for training and testing sets with different numbers of hidden nodes. In the table, RMSE_tra and RMSE_tst denote the performance of training set and testing set, respectively, and the stopping epochs for different hidden layers nodes are identified by bold and italic type. Figure 9.7 presents the results and absolute errors of water levels for the validation data set with the ANN-GA model. A similar procedure is applied to an ANN model. Table 9.9 shows the results comparison between the ANN and ANN-GA models, which indicates that the integration with GA is able to accelerate the convergence of the conventional ANN model. With the same RMSE_vali, the ANN-GA model requires only 135 s whilst the ANN model takes 4,096 s, a more than 30-fold difference.

Table 9.10 summarizes the performance comparison of LR, ANN, GA and ANN-GA models with various measures: RMSE_tes, RMSE_vali, training time, and number of parameters. It can be observed from both RMSE_vali and RMSE_tra that the ANN-GA model, among the various algorithms, performs the best in accuracy. This is because the ANN-GA model possesses the ability to contort itself into a complex form in order to accommodate the temporal changes of the input–output data pairs. This is

Table 9.8 Sensitivity analysis of the numbers of hidden nodes in ANN-GA model

Epochs	Nodes	1	2	3	4	5	6	7
1	RMSE_tra	3.3313	3.5354	3.3792	4.0290	3.1916	4.5184	5.1050
	RMSE_tes	3.3212	3.5325	3.3295	4.0870	3.2609	4.4393	4.9882
50	RMSE_tra	0.2272	0.2914	0.2526	0.2244	0.2384	0.2521	0.2557
	RMSE_tes	0.2880	0.2997	0.2912	0.2838	0.2513	0.2632	0.2607
100	RMSE_tra	0.2192	*0.2197*	0.2234	0.2183	*0.2179*	*0.2189*	*0.2322*
	RMSE_tes	0.2473	*0.2440*	0.2555	0.2751	*0.2483*	*0.2575*	*0.2432*
200	RMSE_tra	0.2184	0.2185	*0.2156*	*0.2150*	0.2146	0.2144	0.2234
	RMSE_tes	0.2458	0.2465	*0.2360*	*0.2525*	0.2604	0.2622	0.2471
300	RMSE_tra	0.2183	0.2183	0.2152	0.2129	0.2137	0.2137	0.2202
	RMSE_tes	0.2458	0.2459	0.2491	0.2754	0.2691	0.2846	0.2465
500	RMSE_tra	0.2182	0.2182	0.2131	0.2121	0.2120	0.2132	0.2125
	RMSE_tes	0.2457	0.2457	0.2727	0.2949	0.2960	0.2965	0.2747
750	RMSE_tra	*0.2182*	0.2182	0.2121	0.2113	0.2106	0.2121	0.2094
	RMSE_tes	*0.2456*	0.2456	0.2987	0.3174	0.2974	0.3032	0.2939
1000	RMSE_tra	0.2182	0.2182	0.2118	0.2108	0.2098	0.2116	0.2089
	RMSE_tes	0.2456	0.2456	0.2963	0.3242	0.2923	0.3139	0.3008

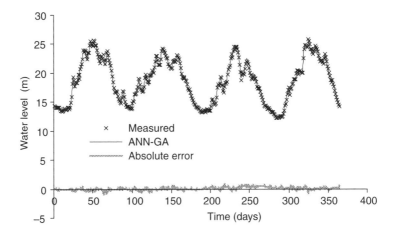

Figure 9.7 Results and absolute errors of water level for validation data set with the ANN-GA model

in contrast to an LR model, which can only fit a linear function to input–output data pairs. It is quite reasonable that an ANN-GA model with 16 parameters can behave more flexibly than an LR model with four parameters. This is analogous to the performance comparison between a power or polynomial function and a simple linear function. Results illustrate that the coupling of GA shortens the training time of the conventional ANN model whilst the integration of ANN improves the local searching capability and

Table 9.9 Comparison of performance between ANN and ANN-GA models

Epochs	ANN			ANN-GA		
	RMSE_tra (m)	RMSE_vali (m)	Training time (s)	RMSE_tra (m)	RMSE_vali (m)	Training time (s)
1	19.610	19.732	0.5	3.379	3.367	113
50	6.174	6.286	4	0.258	0.262	125
100	4.026	4.102	8	0.235	0.241	131
200	2.523	2.532	21	0.213	0.226	135
500	2.013	2.042	52			
1,000	1.814	1.833	89			
1,500	1.691	1.705	139			
2,500	1.516	1.524	199			
5,000	1.184	1.185	445			
10,000	0.664	0.673	600			
20,000	0.401	0.402	1044			
50,000	0.309	0.311	1732			
100,000	0.268	0.272	4096			

Table 9.10 Performance comparison for different models

Model	RMSE_tra(m)	RMSE_vali(m)	Training time(s)	Number of parameters
LR	0.238	0.237		4
ANN	0.268	0.272	4096	16
GA	0.240	0.238	65	4
ANN-GA	0.213	0.226	135	16

hence the accuracy of a traditional GA model. The better performance of the ANN-GA model, in comparison with the ANN or GA model, is justified since the hybrid model is able to take advantage of the local optimization of ANN and the global optimization of GA. It is therefore hoped that the ANN-GA algorithm will have great potential for further developments and applications in future.

9.8   Prototype application III: water stage forecasting by PSO-based ANN

In this case study, PSO is employed to train multilayer perceptrons for river stage forecasting. With this model, real-time water levels in the Shing Mun River of Hong Kong with different lead times are forecast according to the upstream gauging stations or stage/time history at the station itself.

9.8.1   The study site and data

In this case study, the potential flood hazard in the Shing Mun River network, Hong Kong, is investigated. Chau and Lee (1991a, 1991b) and Chau and Chen (2001) furnish more details regarding the location map of the Shing Mun River and its tributaries. Along most of its length, the main conveyance channel is of trapezoidal shape with side slope of 1 in 1.5. The tributaries of the river system comprise three minor streams: the Tin Sam, Fo Tan, and Siu Lek Yuen nullahs. Surface discharge from an extensive catchment area of about 5,200 ha flows into Sha Tin Hoi via the Shing Mun River. The maximum daily runoff is typically less than 5 per cent of the annual flow (Chau and Lee 1991a, 1991b).

In accordance with the daily observed levels both at Fo Tan and at Tin Sam station, located 2 km upstream, water stages at Fo Tan with a lead time of one or two days are predicted. Daily water stages between 1999 and 2001 are used in the study. Data from 1999 to 2000 and data in 2001 are used for training and validation purposes, respectively. The data series chosen for training and validation are analysed carefully to ensure that each data set comprises both high and low discharge periods of the year and also rapid changes in water stages.

In this study, initially two models are developed. In both models, the architecture is 1-3-1: an input layer with one neuron, a hidden layer with three neurons, and output layer with one neuron. The input neuron denotes the water elevation at the current day whilst the output node represents the water elevation after one day or two days, following Thirumalaiah and Deo (1998). This approach has been found to have better results than using two neurons with both a one-day- and two-days-ahead forecast in the output layer. In the training stage, the single input neuron denotes time-series information on water elevations. Moreover, the number of nodes in the hidden layer is determined after a trial and error process during the course of training. A third model with a seven-days-ahead forecast is also tested so as to assess the performance of the model in longer-term forecast.

The stopping criterion is set to be 20,000 training epochs. The sigmoid function is adopted as the transfer function at both the hidden and output nodes. Preprocessing of data is performed first, and all source data are normalized to within zero and one, corresponding to the minimum and maximum variable values over the entire data set, respectively. The PSO parameters adopted are as follows: the number of population is 40; the maximum and minimum velocity values are 0.25 and −0.25, respectively. All these values are acquired through a trial and error procedure.

9.8.2 Results and discussion

In order to gauge the performance of the PSO-based multi-layer ANN, it is evaluated together with a benchmarking BP-based network. For comparison on the same basis, the training process of the BP-based perceptron starts by employing the best initial population of the corresponding PSO-based perceptron. In this study, evaluation of the models is based on three performance criteria: the coefficient of efficiency (R^2), root mean squared error (RMSE) and mean relative error (MRE). The coefficient of efficiency is defined as follows:

$$\text{coefficient of efficiency} = 1 - \text{sum of squared errors divided by total sum of squares}$$

Tables 9.11 and 9.12 present result comparisons for the two different perceptrons using input water level data at Fo Tan station and Tin Sam station, respectively. It can be seen that the PSO-based perceptron performs better in both the training process and the validation process than those by the BP-based perceptron. Moreover, the accuracy of water stage prediction at Fo Tan station made by employing the data gleaned at the upstream station is generally higher compared to its counterparts with the input data gleaned at Fo Tan station itself. This can possibly be justified by the lead time entailed

Table 9.11 Results for forecasting at Fo Tan based on data at the same station

Algorithm	Lead time (days)	Training			Validation		
		Goodness-of-fit Measure					
		R^2	RMSE	MRE	R^2	RMSE	MRE
BP-based	1	0.96	0.16	0.09	0.96	0.21	0.12
	2	0.93	0.24	0.15	0.92	0.29	0.24
	7	0.89	0.35	0.27	0.88	0.43	0.38
PSO-based	1	0.99	0.08	0.04	0.99	0.12	0.06
	2	0.99	0.14	0.07	0.98	0.16	0.09
	7	0.95	0.25	0.18	0.92	0.32	0.21

Table 9.12 Results for forecasting at Fo Tan based on data at Tin Sam (upstream of Fo Tan)

Algorithm	Lead time (days)	Training			Validation		
		Goodness-of-fit Measure					
		R^2	RMSE	MRE	R^2	RMSE	MRE
BP-based	1	0.97	0.14	0.07	0.96	0.16	0.10
	2	0.94	0.21	0.12	0.93	0.24	0.20
	7	0.91	0.30	0.22	0.89	0.41	0.32
PSO-based	1	0.99	0.07	0.04	0.99	0.09	0.05
	2	0.99	0.11	0.06	0.98	0.14	0.08
	7	0.96	0.22	0.16	0.93	0.29	0.18

for flowing from the upstream section to the downstream section and the correlation between the water stages at the two locations.

9.9   Conclusions

In this chapter, three types of evolutionary algorithms – genetic algorithms (GA), genetic programming (GP), and particle swarm optimization (PSO) – are delineated. Three real applications of evolutionary algorithms are also demonstrated.

In the first case study, GP is employed for the analysis of algal dynamics data from a coastal monitoring station in Tolo Harbour, Hong Kong. It is apparent that the interpretation of GP equations is able to identify key input variables that comply with ecological reasoning. It is seen that chlorophyll-a itself is sufficient as input variable to forecast itself at a time lag, indicating an auto-regressive nature of the algal dynamics in the semi-enclosed coastal waters in Tolo Harbour. Results for the forecast of chlorophyll-a indicate that the use of biweekly data can mimic long-term trends of algal biomass reasonably well. However, it might not be appropriate to use these biweekly data for short-term forecast of algal blooms. It is recommended to use input data at a higher frequency in order to enhance the performance of the forecast.

It is shown from the second case study that, when prudent treatment has been taken to prevent over-fitting problems, the ANN-GA model generates accurate flood forecasting of the channel reach between Luo-Shan and Han-Kou stations in the Yangtze River. It is illustrated that this model is capable of avoiding the complication of a conventional mechanistic model, and in particular the requirement to collect a large quantity of site-specific parameters. It couples the advantage of ANN for fast convergence and local optimization with the advantage of GA for global searching capability. It should be noted that the accomplishment of more accurate performance may be in return for additional modelling parameters and possibly larger computation effort when compared with the empirical LR and GA models. Nevertheless, hybrid models such as ANN-GA model are feasible alternatives to conventional models. It is worth exploring different types of hybrid techniques because it might unveil a novel solution approach with more accurate performance.

The third case study is the use of a PSO-based perceptron approach for real-time water level forecasting in the Shing Mun River of Hong Kong with different lead times according to the upstream gauging stations or stage/time history at the station itself. It can be observed from the training and verification period that the water level forecast results are more accurate when compared with the benchmarking BP-based perceptron. Moreover, the accuracy of water stage prediction at Fo Tan station made by employing the data gleaned at the upstream station is generally higher than its counterparts with the input data gleaned at Fo Tan station itself. The initial result shows that the PSO technique can act as an alternative training algorithm for ANNs in water resources applications. More rigorous testing on more complex problems will be undertaken in future works.

A comprehensive investigation into the application of various evolutionary algorithms to coastal modelling has yet to be undertaken, but the early indications of their use in this regard are promising.

10 Knowledge-based systems

10.1 Introduction

In the past decade, the potential of artificial intelligence (AI) techniques for providing assistance in the solution of engineering problems has been recognized. AI deals with the development of cognitive models and computer programs to emulate the intelligence of human beings. In the early work on AI, researchers attempted to develop general problem solvers, which are categorized as weak methods. However, these efforts were met with a number of impediments. Their power was found to be quite limited in so far as solving practical complex problems was concerned. One of the reasons is that they may lead to combinatorial explosion as the complexity of problems increases. Another important reason was that most of the problems these methods solved were common-sense reasoning tasks, i.e. they did not require any special kind of knowledge to solve. It was then suggested that AI techniques could be made more effective by adding domain knowledge. This led to the development of knowledge-based systems (KBS). In this chapter, the characteristics of knowledge-based systems are described. Several real applications of knowledge-based systems are also demonstrated.

10.2 Knowledge-based systems

Knowledge-based systems are defined in a variety of ways by different researchers. In fact, different terminology, namely, "knowledge-based system", "expert system" or "knowledge-based expert system", has been used to represent this type of system. Sriram *et al.* (1985) defined the expert system as that which performs tasks that require a great deal of specialized knowledge that experts in a particular field acquire from long experience with such tasks. Fenves (1989) delineated the system as the practical problem-solving tools that can reach a level of performance comparable to that of a human expert in some specialized problem-solving domains. Dym and Levitt (1991) referred to the system as a computer program that performs a task normally done by an expert or consultant and which, in so doing, uses captured heuristic knowledge. Gaschnig *et al.* (1981)

described a knowledge-based expert system as an interactive computer program incorporating judgement, experience, rules of thumb, intuition, and other expertise to provide knowledgeable advice about a variety of tasks.

The terminology of an expert system entails it to subsume expert knowledge possessed by an expert whereas, as a matter of fact, only a few systems can be claimed to resemble a human expert (Kumar 1995). Adeli (1988) used a less ambitious term for most developed systems as "knowledge-based systems". Nevertheless, they can still be referred to as "expert systems" or "knowledge-based expert systems" owing to their objectives in mimicking the reasoning process and decision-making of human experts. In this chapter, the terminology "knowledge-based system" is used in order to reflect the fact that it uses domain-specific knowledge in the knowledge base, which contributes significantly to the system.

10.2.1　Components of knowledge-based systems

Figure 10.1 shows the organization and operating environment of a typical knowledge-based system, which consists of the following main basic components: knowledge base, inference engine and system context.

1.　**Knowledge base** – The heart and core of any knowledge-based system is the knowledge base, which comprises the crucial problem-solving knowledge in the specific problem domain. The knowledge base is a collection of general facts, documented definitions, well-established

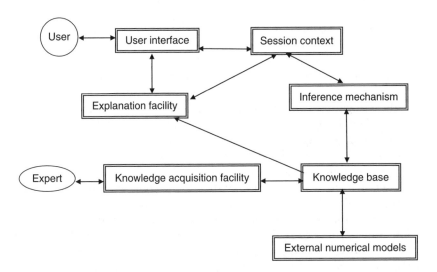

Figure 10.1 Typical components of a KBS system

theory, rules of thumb, heuristic information, judgemental data and causal models of the behaviour specific to the problem domain.

2.   **Inference engine or inference mechanism** – The inference engine, being a knowledge processor that incorporates reasoning methods, monitors the execution of the program by using the knowledge base to modify and manipulate the context. It acts upon the working memory and the knowledge in the knowledge base to solve the stated problem and generate an explanation for the solution. It also determines the problem-solving strategy on sequencing as well as firing of the production rules or procedural methods.

3.   **Working memory or context** – The working memory or context is a workspace for the problem constructed by the inference mechanism from the information provided by the user and the knowledge base. It contains facts as well as all the information that describes the problem being solved, and reflects the current state of the solution process, including both information provided by the user about the problem and the intermediate to final results derived by the system. The organization of the context depends on the nature of the problem domain. The working memory builds up dynamically as a particular problem is being considered: hence its contents change at different stages of the problem. The context is used by the inference mechanism to guide the decision-making process.

Besides the three main modules described above, the system should also be provided with three other components that are not necessarily part of every knowledge-based system but are required to contribute a more functional system. They are described as follows:

1.   **Knowledge acquisition facility** – The knowledge acquisition module, which assists in the translation of knowledge acquired from experts to the required internal format of the system, serves as an interface between the human experts and the knowledge-based system. It provides a means for entering domain-specific knowledge into the knowledge base and revising this knowledge when necessary. It may include a debugging facility, which helps to ensure the correct translation of knowledge into the required format and to check different types of available knowledge representations such as production rule system, procedural method and declarative format.

2.   **User interface** – The function of the user interface module is to accept a problem description from the user and to access and query the system in order to analyse the problem or augment the capability of the system. It provides a friendly interface between the user and the knowledge-based system, usually as a command language for directing execution or in the form of menus, multiple windows, icons or graphics. The interface is responsible for translating the input as specified by the user to the form used by the knowledge-based system, and for handling the

interaction between the user and the knowledge-based system during the decision making process.

3. **Explanation facility** – The explanation module provides the user with explanations of the reasoning inferences used by the knowledge-based system at any point during the consultation session. This explanation can be a priori – why a certain fact is requested – or a posteriori – how a conclusion was reached. It may also contain a help facility, which assists and directs the user to operate the system effectively.

10.2.2 *Characteristics of knowledge-based systems*

The main concept behind knowledge-based systems is to separate the domain-dependent knowledge and the domain-independent control rules to manipulate the knowledge.

Genuine knowledge-based systems may be said to have the following attributes (Adeli 1986; Maher 1987):

- Knowledge-based systems are knowledge-intensive.
- The knowledge for problem-solving is represented principally in symbolic terms rather than numerical terms.
- Knowledge-based systems tend to mimic the decision-making and reasoning processes of human experts in solving a specific complex problem, by providing expert advice, answering questions, and justifying their conclusions.
- Knowledge-based systems can explain the reasoning behind their reaching a particular solution.
- Knowledge-based systems often employ heuristics and rules of thumb as well as compiled knowledge in a specific domain of knowledge to improve the efficiency of search.
- There is usually a separation between the domain knowledge and the methods of manipulating that knowledge in knowledge-based systems.
- Knowledge-based systems have transparent knowledge bases and thus it is usually easy to expand the knowledge bases.
- Domain knowledge is usually divided into many separate independent entities or modules. Knowledge representation and process that employ the knowledge are transparent in order not to be obscured by the implementation language.
- The systems are usually highly interactive.

10.2.3 *Comparisons with conventional programs*

Knowledge-based systems are very much different in nature from conventional algorithmic models. The conventional algorithmic program is constituted by data and program whereas a typical knowledge-based system can be divided into three major parts: an explicit knowledge base, an inference engine and context (Fenves 1989). Also, traditional programs deal with numerical processing whilst knowledge-based systems are involved with symbolic processing. For a conventional program, the order of execution of statements is predetermined. Updates need considerable effort other than by the programmer. The programmer must ensure completeness and uniqueness of the solution. The user, perceiving the program as a black box, has no idea why certain results have been produced. By contrast, knowledge-based systems eliminate the above impediments by partitioning between the knowledge base and the control strategy. This allows for incremental addition of knowledge without manipulation of the overall program structure, and hence the programmer need not guarantee completeness. By ranking several alternatives with inexact inference methods, several solutions with different confidence factors can be provided for a particular input condition. The user can also question the results through the explanation module.

10.2.4   Development process of knowledge-based systems

Conventionally, two types of person are involved during the development of knowledge-based systems: the knowledge engineer and the domain expert. The former is the one who is responsible for acquiring expert knowledge from the domain expert and then transforming it into a knowledge representation format appropriate to that knowledge-based system, whilst the latter possesses the necessary problem-solving heuristics and knowledge for the specific domain problem. However, the recent view is that it is preferable to combine these two separate roles into a single identity. It is believed that a knowledge engineer who is also conversant in the application domain problem is highly desirable, since he or she is able to be fully aware of all the pertinent issues and thus can avoid the potential occurrence of misunderstanding or communication problems with the domain expert. Otherwise, the knowledge engineering process for implementation and development of a knowledge-based application is quite similar in nature to its counterparts of other generic software life cycles, with steps in the development process as follows (Maher *et al.* 1988):

1. **Problem identification** – First of all, the domain problem, together with its nature, overall objectives, domain experts, available resources and computing facilities, needs to be identified.
2. **System architecture** – The overall architecture and framework of the system are established according to various factors, including the reasoning processes employed by the expert in solving the problem, availability of data, problem-solving strategies, information flow of data, etc.
3. **Knowledge acquisition** – This step signifies the process of gleaning the expert domain knowledge by the knowledge engineer from the domain expert.

4. **Development and implementation** – This step consists of structuring the acquired knowledge into the representation format of the adopted development tool, designing an appropriate and user-friendly consultation and user interface, and coding for the control strategy. Together they constitute a prototype or a partial knowledge-based system, which can be operational but not yet verified.

5. **Validation and verification** – After a prototype system has been developed, it should be validated and then verified subject to a wide range of real applications. All syntactic and logical errors in the system are then detected and ratified by expanding or modifying the knowledge employed by the program.

Because knowledge is always being updated, knowledge engineering inevitably deals with an iterative process, and it is always necessary to expand, modify or fine-tune the behaviour of the system to be aligned with the desired result. The increase of depth of knowledge and breadth of the capabilities of the system, and the improvement of the user interfaces and the explanation facilities of the prototype system lead to the enhancement of the overall capability of later versions of the system, compared to earlier ones. However, it is sometimes difficult to demarcate distinctly the steps in the system development, which often overlap. In practice, various generic development processes consisting of a different organization of steps have been suggested, which are determined largely by the preference of the knowledge engineer as well as the nature of the specific domain problem. Another example of the classification of the hierarchy in the system development process is described by Badiru (1992). Under this hierarchical process, maintenance is considered the additional final step after commissioning of the system.

10.2.5 Development tools for knowledge-based systems

One of the most important considerations during the development of knowledge-based systems, which may substantially affect the success of the system, is adoption of the appropriate development tool. Nowadays, a number of development languages as well as tools are available, which can be broadly classified into the following major groups (Dym and Levitt 1991):

1. **General-purpose conventional programming languages** – The conventional programming languages such as Fortran, Pascal and C, which are tailored for algorithmic sequencing, have been used for symbolic programming, but are found to be not too efficient in mimicking human reasoning. Therefore special programming languages designed for artificial intelligence are required to accomplish effective representation

and operation of symbolic processing in the development of knowledge-based systems. Amongst them, the high-level, general-purpose programming languages most commonly used to develop knowledge-based systems are Prolog and Lisp. However, the programming effort required will be tremendous (Adeli 1988).

2. **General-purpose representation tools** – These tools, being programming environments tailor-made for general-purpose knowledge representation, are not limited to a certain type of inference control and strategy. They therefore facilitate the implementation of a range of applications in different fields. However, the main disadvantage of these tools, just like their counterparts in the general-purpose programming languages, is the demanding requirement of the application developers to write and maintain an enormous amount of programming code in order to produce a working knowledge-based system. Nevertheless, they score over general-purpose programming languages in that they furnish better environments, such as powerful database interfaces and friendly user interface utilities for application implementation, and they are implemented on a variety of hardware platforms. Typical examples of some general-purpose tools are OPS5 (Forgy 1981), SRL (Wright and Fox 1983) and UNITS (Stefik 1979). Prolog can also be categorized under general-purpose representation tools because it possesses an inference engine and represents knowledge in declarative knowledge format.

3. **Domain-independent development shells** – Expert system programming environments, often known as expert system shells, have been developed in order to facilitate the implementation of knowledge-based systems. These expert system shells encapsulate specified inference control strategies as well as knowledge representation techniques. They usually provide the system builder with one or more knowledge representation forms and inference mechanisms from which an application can be built by adding domain-specific knowledge. This is very helpful in rapid prototyping of the knowledge-based system. They furnish the framework skeleton of knowledge-based systems, which typically contain the medium of knowledge representation, inference engine and user interface. The application developer is required merely to fill in the domain knowledge and the system can then carry out the requisite function of a functional knowledge-based system with the desired representation technique and inference strategy. Both the number and variety of the commercial expert system shells are increasing rapidly. Some typical examples are Visual Rule Studio (Rule Machines Corporation 1998), AGE (Nii and Aiello 1979), EMYCIN (van Melle 1979), EXPERT (Weiss and Kulikowski 1979), Hearsay-III (Erman *et al.* 1981), Insight 2+ (Level Five Research 1986), Level 5 Object (Information Builders, Inc. 1995), VP-Expert (Friederich and Gargano 1989) and Loops (Bobrow and Stefik 1983).

4. **Special-purpose integrated development environments** – Special-purpose development environments are considered to be of a higher level than the usual expert system development shells since they subsume all the properties of the domain-independent shells together with additional powerful functions such as integrated editors, debugging tools and user interface development facilities. These high-level tools can furnish a single knowledge engineering development environment, which can at the same time integrate the salient features of the artificial intelligence languages appropriately and efficiently. Typical examples of this type of approach are ART (Clayton 1985) and KEE (Intellicorp 1986).

Of course, different classification methods can be employed for these development tools, if all available tools can be covered completely. Adeli (1988) differentiated the environments for the development of knowledge-based systems into three principal groups: the early research tools, the large systems requiring mainframe computers, and the small shells available on microcomputers. Badiru (1992), however, grouped the development environment in terms of networking or independent usage: that is, a single user operating a personal computer or multiple users working simultaneously under the connected engineering workstations.

There exist a wide range of available tools for the development of knowledge-based systems. In addition, owing to the recent advent of artificial intelligence technology, new and more powerful commercial products are emerging at an extremely fast rate together with quickly varying capabilities. However, any of these tools could more or less serve the main purpose of most knowledge-based system applications whilst none of the tools was only tailor-made or could only be applied for a particular purpose (Kumar 1995). Details of the criteria for the selection of the most appropriate knowledge-based system software as well as a survey of these commercial tools are presented in Fazio *et al.* (1988), Sakr and Hosain (1989) and Mohan (1990).

10.2.6 *Knowledge representation*

A number of representation formalisms have been developed to represent complex knowledge structure. The most widely used one is the production rule system model; other forms of representation commonly used are logic, frame-based schemes, semantic networks and, more recently, the object-oriented approach and the blackboard architecture. The capability of these representation schemes to express information on declarative and procedural knowledge differs widely. Most systems employ declarative knowledge representation formalism, which is in fact a collection of relationships between symbols. In the following sections, the conventional rule-based expert systems and the recently popular blackboard architecture are described in more detail.

10.3   Rule-based expert systems

In rule-based representation, the knowledge is encoded in a collection of antecedent-consequent pairs or If–Then rules, and uncertainty in the knowledge is represented by means of confidence factors. If the antecedent of a rule (If statement or statements) is found to be true, then the inference engine may fire the rule, inferring the Then statement or statements. Each rule, which represents an independent chunk of knowledge, is useful for representing the interaction between declarative and procedural knowledge. The production rule system brings the advantages of simplicity and homogeneity, permitting self-examination; the main disadvantage is its limited capability to represent the relationship between various pieces of knowledge. It uses unordered data-sensitive rules as the fundamental unit of computation, which are in stark contrast to the sequenced instructions employed by conventional procedural programs. Production rule systems are most appropriate when the knowledge to be represented occurs naturally in a rule form, where the relationships between rules are extremely complex, and where frequent changes in the knowledge are anticipated.

10.3.1   *Problem-solving strategy*

Broadly speaking, two problem-solving approaches exist, namely, the state-space method and the problem-reduction method. The former searches for a solution in a space of possible solutions whereas the latter first decomposes a problem into a number of sub-problems which then combine together to form the solution to the problem. Although knowledge-based systems are considered strong problem solvers through employing domain knowledge in the solution strategy to reduce the search space, the search technique is still crucial. Commonly used search methods include depth-first search, breadth-first search and best-first search (Fenves 1989). A reasoning or search process is usually carried out in either of two directions, namely, forward chaining and backward chaining. The direction of search is also named control strategy, problem-solving strategy or inference mechanism, with details summarized below.

1.   **Forward chaining** (also referred to as bottom-up, data-driven, forward reasoning or antecedent-driven control strategy) – In this strategy, the system works from an initial state of known facts towards a goal state. Hence this strategy becomes useful in circumstances with a large number of proven hypotheses or goals but with relatively few known input data. A special case of forward-chaining mechanism is known as event-driven, in cases where the problem-solving mechanism is also controlled by the events occurring during the solution process.

2. **Backward chaining** (also termed top-down, goal-driven, backward reasoning or consequent-driven control strategy) – When this strategy is employed, the system attempts to validate a goal or hypothesis by matching all known facts in the system context with the antecedent entailed to fire this hypothesis. This process can also be considered as an inverse order search by the state-space method, commencing in the opposite direction from the goal state to the initial state. The backward-chaining strategy is most efficient when there exist only a very few known goal states. In practice, the backward chaining strategy is often employed in knowledge-based systems of a diagnostic type.
3. **Hybrid chaining** – This strategy is a combination of forward chaining strategy and backward chaining strategy.

All these control strategies together constitute the fundamental search techniques. They may be called by other problem-solving methods and hence become embedded within other methods. Various types of commonly used problem-solving methods exist, include backtracking, constraint handling, plan-generate-test, means-end analysis, hierarchical planning and least commitment principle, agenda control and top-down refinement (Kumar 1995). These methods have so far been applied to solve a range of hypothetical reasoning problems. However, they are often classified as weak methods since the applications of these methods in solving actual complex problems are quite limited.

The production rule system has been the most favourable and versatile representation approach for constructing knowledge-based system. However, it is often found, in particular for complex engineering problems, that using rules alone cannot represent thoroughly all the complex objects and concepts. Model-based reasoning systems, which encapsulate several knowledge representation formalisms including rules, frames and object-oriented programming, emerged. Frames were used extensively to represent the generic components of the engineering problems whilst slots in a frame and their behaviours were inherited by instances of the components included in the engineering system. Because of its modularity, data abstraction and inheritance characteristics, object-oriented programming will probably subsume other approaches in the very near future. Since each knowledge representation scheme has both its pros and cons, the recent trend is to combine various techniques in order to take advantage of the capabilities of each technique to suit the specific domain problem.

10.4 Blackboard architecture

The blackboard architecture is intended to support the development of systems in certain domains characterized by interaction between a multitude of knowledge sources, and hence provides a framework for integrating knowledge from several sources into a single system. Through the integration of rules, frames and object-oriented programming technique, a variety of knowledge sources representing specialized expertise are grouped into individual knowledge modules. The blackboard system encapsulates information-sharing through the common data structure called a blackboard. As shown in Figure 10.2, the blackboard compiles the data entries as well as acting as the sole communication link between various knowledge sources (Hayes-Roth 1983). As such, the blackboard, acting in the role of the global system context, stores the current state of the problem, including problem data, intermediate parameters and final outputs. A common analogy of the blackboard system may be made with problem-solving in domains where a number of experts in different areas of specialism cooperate over the solution which any one of them could never achieve alone. In order to facilitate this process, they agree to use a blackboard to post or write any partial result they can contribute separately. Each expert takes turns to write on the blackboard and, in cases where more than one expert wishes to write simultaneously, the conflict is resolved by some predefined strategy. The blackboard architecture has been used successfully in solving a wide range of tasks, such as speech recognition, signal processing, and planning (Engelmore and Morgan 1988).

The blackboard architecture employs a hierarchical type of knowledge base and opportunistic reasoning whereby several knowledge sources contribute to the reasoning strategy. The blackboard model is often employed to solve complex problems, which are first partitioned into sub-problems. The problem-solving behaviour on the whole relies very much on the interaction of individual processors, known as knowledge modules, since each of these modules undertakes an individual hypothetical solution to the sub-problem at different abstraction levels of the domain problem. The communication between various modules takes place only via the blackboard, which undertakes the key role in integration and development of various hypothetical solutions. A blackboard system consists of a number of knowledge sources communicating via a blackboard and is controlled by an inference mechanism. The major components of a typical blackboard system are the blackboard, knowledge sources, entries and inference mechanism, which are shown as follows (Engelmore and Morgan 1988):

- **The blackboard** represents a global data structure that consists of entries generated by the knowledge sources during the problem solving-process. It is typically partitioned into a number of levels, each of which represents different stages or aspects of the solution process. Normally, knowledge sources are specific to certain levels in the blackboard. The activation of a certain knowledge source depends on the entries generated at certain levels in the blackboard while the actions of the knowledge source modify entries at some other levels. The blackboard has the primary role of keeping track of incremental

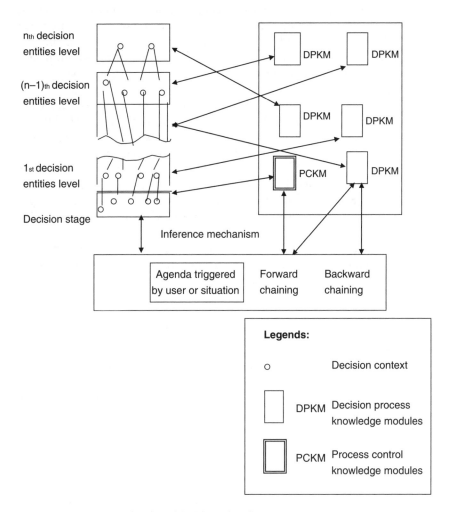

Figure 10.2 Details of the blackboard architecture

changes made in the problem state and thereby of handling the communication among various knowledge sources, until the final solution is attained. The main units in the blackboard are hypotheses, which are either primary guesses about particular aspects of the problem or partial solutions. Hypotheses at various levels are related through structural relationships.

- **Knowledge sources** collectively comprise the knowledge encompassed in the knowledge-based system. They contribute significantly to the creation of entries, which are posted on the blackboard. As an analogy in a production rule system, knowledge sources may be considered equivalent to a collection of production rules. In such an analogy, the antecedent of each rule typically represents the presence or absence of some entries on the blackboard whilst the consequent suggests some actions to be taken leading to some changes to the blackboard entries.

- **Entries** are the immediate results and current state of the solution generated by the knowledge-based system. In a typical system, each entry has a certainty factor as well as a specification.

- **The inference mechanism** usually consists of two main components: the agenda (or scheduler) and the monitor in a typical blackboard model. The agenda keeps track of all the events in the blackboard and calculates the priority of execution for knowledge sources that were generated as a result of the activation of other knowledge sources. It is a list of knowledge sources or rules to be executed in the next cycle. Based on the success or failure of a particular rule, new rules may get added into the knowledge base or some may be deleted from it. The basis of giving priorities to the rules on the agenda may vary from system to system. The monitor takes the element with the highest priority and executes it. Several problem-solving strategies can be implemented using the monitor.

10.5   Advantages and disadvantages of knowledge-based systems

It is necessary to evaluate thoroughly both the advantages and the drawbacks of a knowledge-based system, prior to its adoption and application. In fact, since enormous effort is entailed in developing a functional knowledge-based system, the advantages it offers provide significant justifications for its selection. In this section, the advantages as well as the drawbacks of knowledge-based systems are detailed.

10.5.1   *Advantages of knowledge-based systems*

The most significant advantage of knowledge-based systems is their capacity to solve ill-structured problems such as interpretation and design, which are, by contrast, considered the major shortcomings of conventional algorithmic programs. Also, knowledge-based systems provide an integrated environment for the combination of advanced computer technology together with human expertise knowledge. Many advantages are derived mainly from the separation of the knowledge base from the inference control mechanism (Andriole 1985; Rychener 1988). The major advantages offered by knowledge-based systems are listed as follows:

- Knowledge-based systems provide an effective development environment for software programming, particularly in the engineering field, which often incorporates quite extensively the application of empirical and heuristic knowledge.

- Knowledge-based systems help to convey and distribute human expertise and knowledge from experienced experts to non-experts, through their ability to explain the reasoning process behind the intermediate and final answers.

- Through the establishment of a knowledge-based system, the solution of a domain problem can be standardized and formalized, thus ensuring consistency of the outcome.

- Knowledge-based systems render it possible for the novice user to reach real-time expert-level decisions at relatively low cost. Since they are often implemented in an interactive as well as a decentralized client/server environment, various advantages of popularity of microcomputer facilities should also be taken into account.

- When compared with conventional numerical software, the transparent knowledge base facilitates the extension of knowledge-based systems much more easily, allowing them to enhance their capability gradually as their problem domain evolves or new knowledge is discovered. This in turn gives a good chance for the knowledge engineer and the domain expert to investigate the subtle and complicated areas of the problems, possibly through machine learning.

- Knowledge-based systems may have particular value during consultations for some difficult cases since it can be ensured that, in knowledge-based systems, most of the available data have been utilized whereas human beings may sometimes overlook obscure considerations.

- Knowledge-based systems enhance the chance as well as the consistency of good decision-making, through ensuring objectivity by weighing knowledge without bias and thus minimizing the influence of personal and emotional behaviours of the users.

- Knowledge-based systems free the mind and time of human experts to enable them to concentrate on more creative activities. The ready availability of the system, and its serving as an example of good strategy in approaching a problem, can improve the training environment in industrial settings.

- The modular structure that is commonly adopted in knowledge-based systems leads to dynamic and easy coupling with other programs.

10.5.2 Drawbacks of knowledge-based systems

Successful applications of knowledge-based systems have been found to be those applications that combine facts and heuristics in solving problems to allow merging of human knowledge with computer capability (Dym 1987) whereas those domain problems that are deterministic or numerical in nature are not suitable for knowledge-based systems. Adeli (1988) has summarized some of the drawbacks of knowledge-based systems, when compared with human experts in solving a specific domain problem, as follows.

- The performance of knowledge-based system degenerates very sharply in the vicinity of the boundaries or limits of their expertise knowledge.
- Knowledge-based systems lack common sense and intuition, if this knowledge has not been properly represented in the knowledge base, whilst the human expert may consider them taken for granted.
- It remains a problem to capture rare expertise, which still entails efforts by researchers and domain experts.
- Many knowledge-based systems require expensive and sophisticated artificial intelligence hardware platforms for their normal operation. However, this drawback has been gradually overcome by the recent advances in microcomputer technology.
- Knowledge-based systems currently lack a common user-friendly natural language interface so as to ensure easy operation by non-experts.
- Knowledge-based systems are not strong in solving problems regarding induction or analogy, although they are particularly suitable for problems needing deduction.
- Knowledge-based systems do not know how to learn, if an algorithm for machine learning is not incorporated, whilst the knowledge of the human experts can evolve through their experience.

10.6 Applications and case studies

As a result of years of research in artificial intelligence, knowledge-based systems have emerged covering a wide range of applications in different disciplines. During the last decade, knowledge-based systems have been applied to emulate domain problems owing to their reliability and productivity characteristics (Adeli and Al-Rijleh 1987; Kangari and Boyer 1987; Rouhani and Kangari 1987; Shwe and Adeli 1991; Chau 1992a; Chau and Yang 1994; Chau and Zhang 1995; Chau and Yang 1996; Chau *et al.* 2002; Anuchiracheeva *et al.* 2003; Fdez-Riverola and Corchado 2003; Dai *et al.* 2004; Chau 2006; Pereira and Ebecken 2009; Schories *et al.* 2009). Nevertheless, it has been found that the nature of some problem-solving tasks may render them unsuitable for knowledge-based system formulation. Thus, prior to the implementation of a potential knowledge-based system application, it should be ensured that the following features might be achieved (Adeli and Balasubramanyum 1988).

- **Reliability** – The knowledge-based system must be able to accomplish a high standard of reliability, accuracy and performance covering the whole range of its intended application areas. This requires that the system has the specialized knowledge that distinguishes human experts from novices.
- **Transparency** – In order to allow the system users to thoroughly comprehend the knowledge it encapsulates, the knowledge-based system

must be able to expound explicitly the reasoning mechanism it employs as well as its lines of action.

• **Usefulness** – This will rely highly on the nature of the application domain problem for which the knowledge-based system is developed.

The range of potential knowledge-based system applications covers a spectrum bounded by derivation problems and formation problems at the two ends. In derivation problems, the problem conditions are described as parts of a solution description. This description is completed through applying rules so that the given facts are well integrated into the solution. In formation problems the problem conditions are given in the form of properties that the solution as a whole must satisfy. In real life, most problems fall between these two extreme categories. The following classes of problems are normally encountered at the derivation end of the spectrum: interpretation, diagnosis and monitoring. Formation problems are usually examples of the generate-and-test paradigm, i.e. a possible candidate solution is generated by one part of the system and is then tested for suitability by another part of the system. Formation problems fall into two subclasses: constraint satisfaction and optimization problems. In constraint satisfaction, it is required only that the solution satisfy a set of constraints, while in optimization an attempt is made to find the optimal solution. Planning and design are classes of problems usually encountered at this end of the spectrum.

Areas of early applications of knowledge-based systems technology include medical diagnosis, mineral exploration, chemical spectroscopy, design, data interpretation, planning, and education. All these applications of knowledge-based system can be broadly classified into the following categories:

1. **Design** – Design is the process of developing an appropriate configuration for an object that can satisfy all application constraints. Some typical examples are:

   • SEAWALL is an expert system for the design of gravity-type vertical seawalls by employing a shell Vp-Expert, together with external executable programs written in Turbo Pascal (Chau 1992).
   • THRUSTBLOCK is a knowledge-based expert system on design of thrust blocks for water pipelines (Chau and Ng 1996).
   • R1 is a design system employed to configure VAX-11/780 computer components for Digital Equipment Corporation in accordance with the requests made by the customers (McDermott 1980).

2. **Diagnosis** – Diagnosis is the process of inferring a malfunction situation through observed irregularities and data interpretation. This is the

application area to which knowledge-based system technology has been most successfully applied. Some typical examples are shown as follows:

- CADUCEUS was designed for the diagnosis of diseases on internal medicine (Pople 1982).
- MYCIN was developed to assist physicians in the diagnosis and treatment of meningitis as well as bacteraemia infections (Shortliffe 1976).

3. **Interpretation** – The interpretation process consists of observing data and explaining the meaning through inferring the corresponding problem state. Some typical examples are shown as follows:

- DENDRAL is a knowledge-based system for performing spectroscopic analysis of an unknown molecule and hence for prediction of its molecular structure (Lindsay *et al.* 1980).
- Dipmeter Adviser is a knowledge-based system for interpretation of the log data in a geophysical oil well (Davis *et al.* 1981).
- PROSPECTOR is a knowledge-based system for the identification of the geological ore-bearing formations (Duda *et al.* 1979).

4. **Monitoring** – The purpose of monitoring tasks is to make comparison of observations with the standard in order to plan for vulnerability. A typical example is:

- Ventilation Manager is a knowledge-based system for monitoring the ventilation therapy of a patient (Fagan *et al.* 1979).

5. **Planning** – Planning is a pre-arranged process that results in a set of actions aimed to generate an anticipated outcome. Some typical examples are:

- COASTAL_WATER is an integrated knowledge-based system as a tool of knowledge transfer for personnel on water resources planning and management in coastal waters (Chau 2004d).
- ONTOLOGY_KMS is an ontology-based knowledge management system (KMS) for simulating human expertise during problem-solving by incorporating artificial intelligence and coupling various descriptive knowledge, procedural knowledge and reasoning knowledge involved in the coastal hydraulic and transport processes (Chau 2007).
- RIVER_NET is an integrated knowledge-based system for fluvial hydrodynamics, which couples the synthetic utilization of computer-aided design and artificial intelligence techniques (Chau and Yang 1992; Chau and Yang 1993).

- RUNOFF is a knowledge-based system for flow routing in a river network, which assists in decision-making on selection of numerical models and parameters (Chau and Zhang 1995).
- MOLGEN is a knowledge-based system for planning experiments on molecular genetics (Stefik 1981).

10.6.1 COASTAL_WATER

This prototype KBS on numerical modelling of flow and/or water quality in coastal waters was written using the shell Visual Rule Studio, which is a hybrid application development tool that integrates object-oriented techniques and expert system technology with traditional and procedural programming. Visual Rule Studio installs as an integral part of Microsoft Visual Basic 6.0 as a form of ActiveX Control (in the Visual Basic development language), which is a type of ActiveX file type with the extension .ocx. Moreover, the shell is able to treat not only symbolic problems, but also numerical and graphical problems, which is a characteristic of this problem. The basic structure of an object consists of name, properties, and attributes. The attributes consist of name, type, facets, method, rules and demons. Figure 10.3 shows the structure of a Visual Rule Studio object.

The steps and elements of numerical modelling of flow and/or water quality in coastal waters are carefully examined and, on the basis of this investigation, the main steps of the simulation procedures are then categorized. A database system for the modelling of coastal waters, with particular emphasis on Hong Kong situations, is developed. A knowledge base is developed, merged and then interfaced with the conventional number-crunching

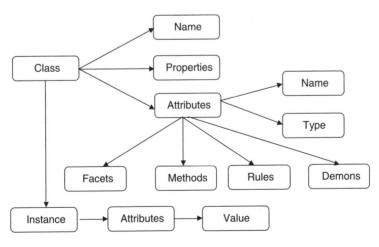

Figure 10.3 Structure of a Visual Rule Studio object

type of analysis and simulation programs to form an integrated automatic modelling of coastal waters incorporated with the facilities of expert advice.

The architecture of COASTAL_WATER is shown in Figure 10.4. Besides the usual components in a typical expert system, namely, knowledge base, session context, inference mechanism, user interface module, knowledge acquisition module and explanation module, it also incorporates executable numerical models. In this prototype system, the knowledge base contains the knowledge on numerical modelling of flow and/or water quality from the

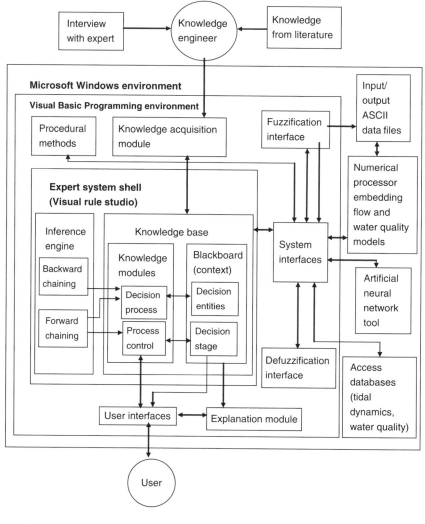

Figure 10.4 Schematic diagram of COASTAL_WATER

literature and domain experts. Since it is a complex task to select a model, the main aim of the prototype system is to establish the knowledge rules based on the analysis of these conditions. Through the operation of these rules, the most effective model can be chosen with respect to accuracy and computational efficiency according to the types and tasks of project.

The knowledge base incorporates the whole set of inference rules relating to the manipulation direction and the user's requirements. The following example gives a typical production rule, which incorporates the fuzzy description:

RULE to determine the model dimensions: 3 of 9
IF the water depth is very deep
AND
the density stratification in the vertical direction is not significant
THEN
the two-dimensional horizontal numerical model is selected with a confidence factor of 80

The IF clause of the above rule statement describes the premises that the water depth should be very deep and that the density stratification should not be significant in order for the conclusion to be fulfilled and hence for the rule to be triggered. The THEN statement of the rule gives the conclusion that the two-dimensional horizontal numerical model is selected with a confidence factor 80. The confidence factor is employed as the determining factor to control the inference process and the selection of each parameter. Its range is basically from 0 to 100, representing the degree of confidence with which the statement is known. The confidence factors are set by experts based on heuristics and experience. The rule base is designed to provide the link between the specifications made by the user and the recommended parameter selection by matching the highest confidence factor. In the above example rule statement, the water depth and the vertical density difference are expressed as a fuzzy description. As an alternative method, the user can also enter their exact numerical values during the query process. The system can then transform the numerical values into the corresponding fuzzy description by a fuzzy member curve, which computes the pertinent confidence of membership prior to searching the rule base for conclusions.

Several commonly used numerical models that have been successfully applied to solve practical engineering problems are incorporated into this prototype system. The numerical models, which can be executed to generate the numerical simulation of real phenomena, are the central component of traditional numerical modelling systems. These well-proven and validated models have often been developed in conventional languages such as Fortran. These external models implement the numerical analysis and result representation tasks. Visual Rule Studio provides the function of causing external files to execute. After the external program has finished its execution, control is returned to the knowledge-base file from which it was called under the Visual Rule Studio environment.

In order to develop a practical engineering KBS, the main task is to create the domain knowledge bases utilizing the built-in reasoning mechanisms of Visual Rule Studio. Reasoning in Visual Rule Studio can proceed in two ways: "forward chaining", which emphasizes the premises of rules; and "backward chaining", which focuses on the conclusions of rules. The inference engine is the mechanism used to solve problems, i.e. to find goals. The inference engine accomplishes this by managing and manipulating the rule base. In order to reach its goal, Visual Rule Studio's inference engine systematically searches for new values to assign to appropriate variables that are present in the knowledge bases. Thus it has the capability to add knowledge to the knowledge base.

The inference engine begins to search for a value to be assigned to its associated goal. If the value assigned to that goal is not known, the inference engine searches the rule base for an assigned value. Initially, not all values of variable that might aid in the search are known. Thus the search must systematically accumulate new knowledge by considering rules from the rule base which might yield helpful facts in the process of finding the goal. Visual Rule Studio does this via a backward chaining method. The effective use of Visual Rule Studio's control strategy when designing the system can create tremendously complicated reasoning paths to simulate the heuristic decision-making involved in the domain experts' engineering routines.

In the development of the prototype system, it was designed with a view that warnings would be issued when necessary and extra text would be provided to help the user answer the questions. It would be very instructive to see immediately the rules that lead to the result for recommended numerical techniques or model parameters. The user interface is implemented by several rules in the knowledge base. The consultation screens are designed to help the users to grasp the physical idea of water resources processes. Figure 10.5 shows a sample screen displaying a graphical user interface incorporated into numerical processing.

When the inference engine has reasoned about the knowledge bases, it calls the numerical modelling program to load and execute the input data files which are generated in the knowledge base. During the running of the numerical model, the output data are saved into files. The output of the numerical modelling program includes the variation of tidal levels, velocities, water quality variables, etc. They can then be processed by Microsoft Excel and translated into graphs, which are convenient for engineers to evaluate. This system has been verified and validated by applying it to several real prototype problems in Hong Kong's coastal waters. The application case studies involve the establishment of several numerical models on coastal flow and water quality in Hong Kong, which encapsulate a few

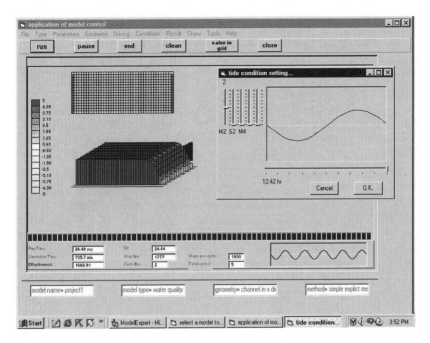

Figure 10.5 Sample screen displaying graphical user interface incorporated into numerical processing

strategically chosen locations such as the Pearl River Estuary. The areas around the Pearl River Estuary have been prospering during the past decade at such a dramatic rate it has resulted in a worsening and deteriorating water quality. Figure 10.6 shows a sample screen showing an interactive graphical display of the topography at the Pearl River Estuary.

10.6.2   ONTOLOGY_KMS

In this work, the layered KM system is built with The KArlsruhe ONtology and Semantic Web (KAON), which is an ontology development environment (University of Karlsruhe 2009). Its characteristics include the open source and distributed component-based J2EE architecture (Sun Microsystems Inc. 2009). The ontology-based KMS therefore has the advantages of high flexibility and robustness. The ontology, comprising concepts, properties, and instances, is grouped into reusable ontology-instance models (Motik *et al.* 2002). Under this development environment, a property may be featured as symmetric, transitive or inverse with other concepts, which has the capability to support a lightweight inference mechanism. In this way, the ontology furnishes a search engine with the functionality of semantic match in a KM system.

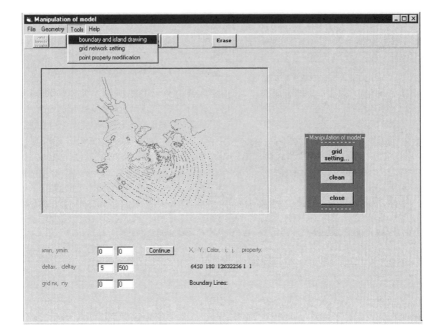

Figure 10.6 Sample screen showing an interactive graphical display of the topography at the Pearl River Estuary

The building of an ontology is often recognized to be the first basic step in facilitating KM activities (Guarino 1997). Figure 10.7 presents the framework of an ontology-based KMS on flow and water quality modelling, in which a three-level architecture for intelligent decision support is adopted. This comprises the application level, the description level and the object level, which are listed in descending order. It should be noted that ontologies are identified in the description level and, through this arrangement, users in the application level are able to access the object-level sources in an intelligent manner. Diverse knowledge sources and information, under the format of numerical data, text streams, validated models, meta-models, movie clips or animation sequences, and so on, collectively termed knowledge objects (KOs), are included in the object level (Nemati *et al.* 2002).

In this work, the ontology is divided into two groups, namely, the information ontology and the domain ontology (Abecker *et al.* 1998). The information ontology represents a meta-model comprising generic concepts and attributes of KOs, which are represented by the Dublin Core (Dublin Core Metadata Initiative 2009). On the other hand, key concepts, attributes, instances and relations of flow and water quality modelling are located in the domain ontology, whose principal role is to attain the functionality of semantic match during the search of KOs. Figure 10.8 shows part of the

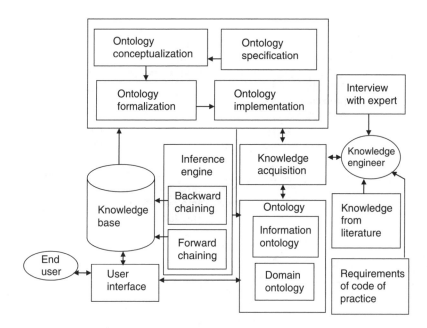

Figure 10.7 Framework of ONTOLOGY_KMS

domain ontology of flow and water quality modelling. It can be observed that there exist various forms of relations, namely, the inheritance relations, functional relations, structural relations, behaviour relations and so on.

During the manipulation stage, when an end-user accesses the knowledge base, the ontology can support tasks of KM as well as searching. The knowledge base and the ontology are linked to one another via both ontology formalization and ontology implementation, which furnishes a route for the extension of the information ontology. During the maintenance stage, knowledge engineers or domain experts can add, update, revise and delete the information ontology or the domain ontology via a knowledge acquisition module.

One of the most difficult issues in flow and water quality modelling is how to select an appropriate model together with the associated parameters. The KM system is able to represent knowledge in a fashion that is appropriate for the modelling of application decision knowledge, to isolate the policies and decisions from application logic and to supply the intelligent support during problem-solving by visual window interfaces. The KMS is tailored so that the application rules are isolated into identifiable and reusable components.

In order to demonstrate the application of the prototype KMS, a case study on the eutrophication problem in Tolo Harbour of Hong Kong is presented. The study area is a nearly land-locked embayment with a narrow

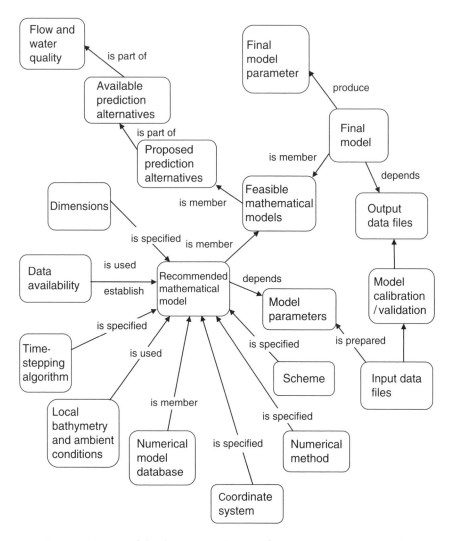

Figure 10.8 Part of the domain ontology on flow and water quality modelling

outlet connecting with Mirs Bay – one of the major south-facing bays in the South China Sea. The water depth varies from about 2 m in the inner part to over 20 m in the outer part of Tolo Channel and about 12 m on average. The average diurnal tidal difference is about 0.97 m, mean high tide is 1.75 m, and mean low tide is 0.78 m. For most of the year, little freshwater is discharged into the harbour, and it could be considered as an embayment. During the summer, however, the differences in surface and bottom water temperature and salinity, caused by solar radiation and rainfall, result in an obviously lighter surface layer and definite mesolimnion in the water column – a two-layered system. Density stratification weakens the vertical mixing and may remove the connection between benthic grazers and near-surface biomass by inhibiting vertical transport. In winter, higher dissolved oxygen levels in the bottom waters are generally recorded due to increased turbulent mixing within the water body, resulting from the strong north-east monsoon. However, during the summer, fewer bottom waters suffer from serious oxygen depletion, even approaching anoxic status, although the dissolved oxygen content in most of the surface was commonly found to be at satisfactory levels, even at super-saturation. Thus, it is necessary to simulate unsteady water quality transport in a density-stratified natural water body. The readers are referred to Chau and Jin (1998) for details of the eutrophication modelling.

The questionnaires are first entered through the user interfaces based on the background of the eutrophication problem for Tolo Harbour in Hong Kong. After the input data have been entered, a summary of the input requirements is shown in the left frame of questionnaires as shown in Figure 10.9. When the command button INFER is clicked, the process of model selection can be automatically attained on the basis of Rule Sets I and Rule Sets II. The right frame shows the inference result about the features of the suggested model for this example, which has been verified to be consistent with the decisions reached by several domain experts.

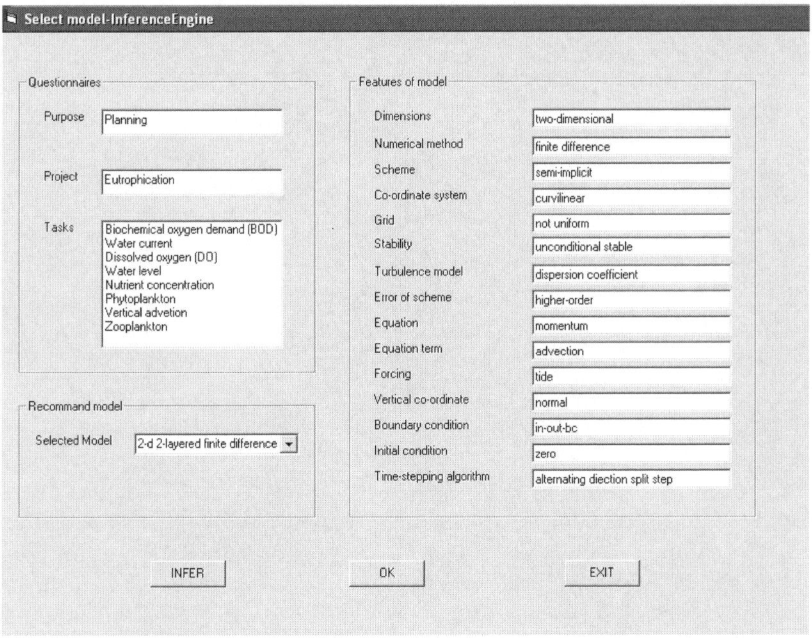

Figure 10.9 Display screen of model selection for the application example

10.7   Conclusions

In this chapter, the characteristics of knowledge-based systems (KBSs) are presented. The applications of KBSs in coastal hydraulics and engineering are discussed. As a real application case study, an integrated prototype KBS, which assists in making decisions on the modelling process of flow and water quality, was developed and implemented. It incorporates an ANN for training of water quality parameters and a fuzzy rule base for representation of the heuristic knowledge. It is shown that the hybrid application of these latest AI technologies is appropriate to act as a repository for heuristic knowledge. The knowledge base is transparent and can easily be updated, which renders the prototype KBS an ideal tool for incremental programming. The prototype system has been successfully calibrated in Hong Kong conditions. This prototype system, serving both as a design aid and as a training tool, is able to allow hydraulic engineers and environmental engineers to become acquainted with up-to-date flow and water quality simulation tools. Moreover, the system can quickly assist policymakers in reaching decisions and also furnish a convenient and open information service on water quality for the general public.

11 Conclusions

Contemporary coastal models are inevitably highly specialized, involve certain assumptions and/or limitations, and are operable only by experienced engineers who acquire a thorough understanding of the underlying theories. This has resulted in significant constraints on the use of models, thus producing a discrepancy between the developers and users of models. These models are usually not user-friendly enough. They lack the capacity to transfer knowledge in the application and interpretation of the model, in furnishing expert backup to novice users, and in accomplishing effective communication from developers to users. Many users of a model do not have the specific knowledge to glean their input data, establish algorithmic models and assess the results of their model. The result may be the outcome of inferior designs and the under-utilization or even total failure of these models. Recently, there has been increased need for an integrated approach. Thus, the problem is to present the information, knowledge, and experience in a format that facilitates understanding by a broad range of users from novices to experts (Abbott 1986).

During the past two decades, the information revolution has fundamentally changed the traditional planning, modelling, and decision-making methodologies of water-related technologies. The recent advances in artificial intelligence (AI) technologies are making it possible to integrate machine learning capabilities into numerical modelling systems so as to bridge the gaps and lessen the burdens on human experts. Information technology now takes a significant role in the sustainable development and management of water resources. In addition, the general availability of sophisticated personal computers with ever-expanding capabilities has generated increasing complexity in terms of computational ability in the storage, retrieval and manipulation of information flow.

This book has reviewed the state-of-the-art of contemporary numerical modelling and progress in the integration of AI into coastal modelling. Attempts to integrate various AI technologies, including knowledge-based systems (KBSs) (Chau *et al.* 2002; Schories *et al.* 2009), genetic algorithms

(GAs) (Chau 2002; Pinthong *et al.* 2009), genetic programming (GP) (Kalra and Deo 2007; Muttil and Chau 2007), artificial neural networks (ANNs) (Recknagel *et al.* 1997; Chau and Cheng 2002) and fuzzy inference systems (Maier *et al.* 2001; Cheng *et al.* 2002) into numerical modelling systems have been detailed and discussed. KBSs have apparent advantages over the others in facilitating more transparent transfers of knowledge in the use of models and in providing the intelligent manipulation of calibration parameters. KBSs may furnish meaningful advice to inexperienced engineers on how to establish a numerical model, although they still need to have an understanding of the problem domain. Of course, the other AI methods also have their individual contributions towards accurate and reliable coastal predictions. With the increase in capability of AI technologies, it is believed that the resulting tool might be very powerful, in view of the possible coupling of the advantages of each technique (Chau 2006a).

To date, individual applications of these innovative AI techniques have been recorded in the literature. However, they were usually adopted for specific situations in an isolated manner. Since the application of different AI technologies is not mutually exclusive, one of the promising directions is the hybrid combination of two or more of the methods as mentioned above to generate an even more versatile coastal modelling system. For example, the use of a hybrid algorithm integrating KBS and ANN is feasible for establishing rules in the KBS on the basis of implicit relationships derived from the ANN. In fact, there is a great deal of potential in extracting the knowledge embedded in the connection weights of trained ANN models, as well as in the highly transparent knowledge representation paradigm of KBS. Similarly, it is also feasible to use GAs to locate the global optimization in ANNs as well as the fuzzy representation of rule sets in KBSs. It is strongly trusted that the integration of AI modules will enhance the applicability of modelling systems in real practice.

Wu *et al.* (2009) endeavoured to couple three data-preprocessing techniques – moving average (MA), singular spectrum analysis (SSA) and wavelet multi-resolution analysis (WMRA) – with an ANN in order to enhance the estimate of daily flows. Six models, including the original ANN model without data preprocessing, ANN-MA, ANN-SSA1, ANN-SSA2, ANN-WMRA1, and ANN-WMRA2, were developed and assessed. The ANN-MA, ANN-SSA1, ANN-SSA2, ANN-WMRA1 and ANN-WMRA2 were developed by employing the original ANN model coupled with MA, SSA and WMRA, respectively. Two different means were used for SSA and WMRA. The models were applied to two daily flow series in two watersheds in China, Lushui and Daning, for three different prediction horizons, namely, one-, two-, and three-day-ahead forecasting. Results indicated that, among the six models, the ANN-MA has the highest accuracy and is able to eradicate the lag effect. It was also noted that the performances from the different means used for SSA and WMRA did not affect the results. Moreover, in that case study, the models based on the SSA performed better than their counterparts of the WMRA at all forecasting horizons. It indicated that the SSA was more effective than the WMRA in enhancing the ANN performance.

Last but not least, it is observed that most of the above studies have been undertaken for fresh water riverine systems and applications to coastal systems have been very scarce. More works can be undertaken to find applications of AI in this area to a fuller extent.

References

Abbott, M.B. (1989). "Review of recent developments in coastal modelling", in R.A. Falconer, P. Goodwin, and R.G.S. Matthew, eds., *Hydraulic and Environmental Modelling of Coastal, Estuarine and River Waters*, Aldershot, Gower Technical, pp. 3–39.

Abbott, M.B. (1991). *Hydroinformatics: Information Technology and the Aquatic Environment*, Aldershot/Brookfield, Avebury Technical.

Abbott, M.B. (1993). "The electronic encapsulation of knowledge in hydraulics, hydrology and water resources", *Advances in Water Resources*, 16(1): 21–39.

Abderrezzak, K.E. and Paquier, A. (2009). "One-dimensional numerical modeling of sediment transport and bed deformation in open channels", *Water Resources Research*, 45, W05404.

Abd-el-Malek, M.B. and Helal, M.M. (2009). "Application of a fractional steps method for the numerical solution of the two-dimensional modeling of the Lake Mariut", *Applied Mathematical Modelling*, 33(2): 822–834.

Abecker, A., Bernardi, A., Hinkelmann, K., Kühn, O., and Sintek, M. (1998). "Toward a technology for organizational memories", *IEEE Intelligent Systems & their Applications*, 13(3): 40–48.

Abualtayef, M., Kuroiwa, M., Tanaka, K., Matsubara, Y., and Nakahira, J. (2008). "Three-dimensional hydrostatic modeling of a bay coastal area", *Journal of Marine Science and Technology*, 13(1): 40–49.

Adeli, H. (1986). "Artificial intelligence in structural engineering", *Engineering Analysis*, 3(3): 154–160.

Adeli, H. (1988). "AI techniques and the development of expert systems", in H. Adeli, ed., *Expert System in Construction and Structural Engineering*, London, Chapman & Hall, pp. 13–20.

Adeli, H. and Al-Rijleh, M.M. (1987). "A knowledge-based expert system for the design of roof trusses", *Microcomputers in Civil Engineering*, 2(3): 179–195.

Adeli, H. and Balasubramanyum, K.V. (1988). *Expert Systems for Structural Design: A New Generation*, Englewood Cliffs, NJ, Prentice Hall.

Alho, P. and Aaltonen, J. (2008). "Comparing a 1D hydraulic model with a 2D hydraulic model for the simulation of extreme glacial outburst floods", *Hydrological Processes*, 22(10): 1537–1547.

Andriole, S.J. (1985). *Applications in Artificial Intelligence*, Princeton, NJ, Petrocelli Books.

Anuchiracheeva, S., Demaine, H., Shivakoti, G.P., and Ruddle, K. (2003). "Systematizing local knowledge using GIS: Fisheries management in Bang Saphan Bay, Thailand", *Ocean and Coastal Management*, 46(11–12): 1049–1068.

Aoki, K. and Isobe, A. (2007). "Application of finite volume coastal ocean model to hindcasting the wind-induced sea-level variation in Fukuoka bay", *Journal of Oceanography*, 63(2): 333–339.

Arhonditsis, G.B. and Brett, M.T. (2005). "Eutrophication model for Lake Washington (USA): Part II – Model calibration and system dynamics analysis", *Ecological Modelling*, 187(2–3): 179–200.

ASCE Task Committee. (2000a). "Artificial neural networks in hydrology – I: Preliminary concepts", *Journal of Hydrologic Engineering, ASCE*, 5(2): 115–123.

ASCE Task Committee. (2000b). "Artificial neural networks in hydrology – II: Hydrological applications", *Journal of Hydrologic Engineering, ASCE*, 5(2): 124–137.

Ataie-Ashtiani, B. (2007). "MODSharp: Regional-scale numerical model for quantifying groundwater flux and contaminant discharge into the coastal zone", *Environmental Modelling and Software*, 22(9): 1307–1315.

Ataie-Ashtiani, B. and Farhadi, L. (2006). "A stable moving-particle semi-implicit method for free surface flows", *Fluid Dynamics Research*, 38(4): 241–256.

Babovic, V. and Abbott, M. B. (1997). "The evolution of equations from hydraulic data, part I: Theory", *Journal of Hydraulic Research*, 35(3): 397–410.

Babovic, V. and Keijzer, M. (2000). "Genetic programming as a model induction engine", *Journal of Hydroinformatics*, 2(1): 35–60.

Babovic, V. and Larsen, L.C., eds. (1998). *Proceedings of the Third International Conference on Hydroinformatics*, 24–26 August 1998, Copenhagen, Denmark, A.A. Balkema.

Badiru, A.B. (1992). *Expert Systems Applications in Engineering and Manufacturing*, Englewood Cliffs, NJ, Prentice Hall.

Baird, J.I. and Whitelaw, K. (1992). "Water quality aspects of estuary modelling", in R.A. Falconer, ed., *Water Quality Modelling*, Aldershot, Ashgate, pp. 119–126.

Baker A.J. (1973). "Finite element solution algorithm for viscous incompressible fluid dynamics", *International Journal of Numerical Methods in Engineering*, 6: 89–101.

Baker, J.E. (1985). "Adaptive selection methods for genetic algorithms", *Proceedings of the 1st International Conference on Genetic Algorithms and their Applications*, J.J. Grefenstette, ed., Hillsdale, NJ, Lawrence Erlbaum Associates, pp. 101–111.

Banzhaf, W., Nordin, P., Keller, R.E., and Francone, F.D. (1998). *Genetic Programming, an Introduction: On the Automatic Evolution of Computer Programs and its Applications*, San Francisco, Morgan Kaufmann.

Barciela, R.M., Garcia, E., and Fernandez, E. (1999). "Modelling primary production in a coastal embayment affected by upwelling using dynamic ecosystem models and artificial neural networks", *Ecological Modelling*, 120(2–3): 199–211.

Bastarache, D., El-Jabi, N., Turkham, N., and Clair, T.A. (1997). "Predicting conductivity and acidity for small streams using neural networks", *Canadian Journal of Civil Engineering*, 24(6): 1030–1039.

Batchelor, G.K. (1967). *An Introduction to Fluid Dynamics*, Cambridge, Cambridge University Press.

Benque, J.P., Haugnel, A., and Viollet, P.L. (1982). "Numerical methods in environmental fluid mechanics", in M.B. Abbott and J.A. Cunge, eds., *Engineering Applications of Computational Hydraulics*, Vol. II, Boston, MA, Pitman.

Bingham, H.B. and Agnon, Y. (2005). "A Fourier-Boussinesq method for nonlinear water waves", *European Journal of Mechanics B: Fluids*, 24(2): 255–274.

Blumberg, A.F. and Mellor, G.L. (1980). "A coastal ocean numerical model", in J. Sunderman and K.-P. Holtz, *Mathematical Modelling of Estuarine Physics, Proc. Int. Symp., Hamburg, Aug. 1978*, Berlin, Springer-Verlag, pp. 203–214.

Blumberg, A.F. and Mellor, G. (1987). "A description of a three-dimensional coastal ocean circulation model", in N.S. Heaps, ed., *Three-Dimensional Coastal Ocean Models*, Washington, DC, American Geophysical Union, pp. 1–16.

Blumberg, A.F., Khan, L.A. and St. John, P. (1999). "Three-dimensional hydrodynamic model of New York Harbor region", *Journal of Hydraulic Engineering, ASCE*, 125(8): 799–816.

Bobbin, J. and Recknagel, F. (2001a). "Inducing explanatory rules for the prediction of algal blooms by genetic algorithms", *Environment International*, 27(2–3): 237–242.

Bobbin, J. and Recknagel, F. (2001b). "Knowledge discovery for prediction and explanation of blue-green algal dynamics in lakes by evolutionary algorithms", *Ecological Modelling*, 146(1–3): 253–262.

Bobrow, D.G. and Stefik, M.J. (1983). *The LOOPS Manual*, Technical Report, Palo Alto, CA, Xerox Corp.

Box, G.E.P. and Jenkins, G.M. (1976). *Time Series Analysis Forecasting and Control*, San Francisco, Holden-Day.

Broom, M.J. and Ng, A.K.M. (1995). *Water Quality Hong Kong and the Influence of the Pearl River, Coastal Infrastructure Development in Hong Kong: A Review*, Hong Kong Government, Hong Kong.

Buonaiuto F.S. Jr. and Bokuniewicz, H.J. (2008). "Hydrodynamic partitioning of a mixed energy tidal inlet", *Journal of Coastal Research*, 24(5): 1339–1348.

Carballo, R., Iglesias, G., and Castro, A. (2009). "Residual circulation in the Ria de Muros (NW Spain): A 3D numerical model study", *Journal of Marine Systems*, 75(1–2): 116–130.

Carter, H. and Okubo, A. (1965). *A Study of the Physical Processes of Movement and Dispersion in the Cape Kennedy Area: Final Report under the US Atomic Energy Commission Contract No. AT(30-1)-2973*, Chesapeake Bay Institute, The John Hopkins University.

Casulli, V. and Cheng, R.T. (1992). "Semi-implicit finite difference methods for three-dimensional shallow water flow", *International Journal for Numerical Methods in Fluids*, 15(6): 629–648.

Chan, B.S.S. and Hodgkiss, I.J. (1987). "Phytoplankton productivity in Tolo Harbour", *Asian Marine Biology*, 4: 79–90.

Chang, N.B., Chen, H.W., and Ning, S.K. (2001). "Identification of river water quality using the fuzzy synthetic evaluation approach", *Journal of Environmental Management*, 63(3): 293–305.

Charhate, S.B., Deo, M.C., and Sanil Kumar, V. (2007). "Soft and hard computing approaches for real-time prediction of currents in a tide-dominated coastal area", *Proceedings of the Institution of Mechanical Engineers Part M: Journal of Engineering for the Maritime Environment*, 221(4): 147–165.

Chau, K.W. (1992a). "An expert system for the design of gravity-type vertical seawalls", *Engineering Applications of Artificial Intelligence*, 5(4): 363–367.

Chau, K.W. (1992b). "Two-dimensional characteristics finite element schemes for advection-dominated flow problems", in W.R. Blain and E. Cabrera, eds., *Hydraulic Engineering Software IV: Computer Techniques and Applications*, Southampton, Boston, CMP Elsevier, pp. 123–132.

Chau, K.W. (2002). "Calibration of flow and water quality modeling using genetic algorithm", *Lecture Notes in Artificial Intelligence*, 2557, 720–720.

Chau, K.W. (2003). "Manipulation of numerical coastal flow and water quality models", *Environmental Modelling and Software*, 18(2): 99–108.

Chau, K.W. (2004a). "A three-dimensional eutrophication modeling in Tolo Harbour", *Applied Mathematical Modelling*, 28(9): 849–861.

Chau, K.W. (2004b). "River stage forecasting with particle swarm optimization", *Lecture Notes in Computer Science*, 3029: 1166–1173.

Chau, K.W. (2004c). "Rainfall-runoff correlation with particle swarm optimization algorithm", *Lecture Notes in Computer Science*, 3174: 970–975.

Chau, K.W. (2004d). "Knowledge-based system on water-resources management in coastal waters", *Water and Environment Journal*, 18(1): 25–28.

Chau, K.W. (2005). "A split-step PSO algorithm in prediction of water quality pollution", *Lecture Notes in Computer Science*, 3498: 1034–1039.

Chau, K.W. (2006a). "A review on the integration of artificial intelligence into coastal modeling", *Journal of Environmental Management*, 80(1): 47–57.

Chau, K.W. (2006b). "Development of an integrated knowledge-based system on flow and water quality in Hong Kong coastal waters", *International Journal of Environment and Pollution*, 28(3–4): 297–309.

Chau, K.W. (2007). "An ontology-based knowledge management system for flow and water quality modeling", *Advances in Engineering Software*, 38(3): 172–181.

Chau, K.W. and Albermani, F. (2003). "Knowledge-based system on optimum design of liquid retaining structures with genetic algorithms", *Journal of Structural Engineering, ASCE*, 129(10): 1312–1321.

Chau, K.W. and Chen, W. (2001). "A fifth generation numerical modelling system in coastal zone", *Applied Mathematical Modelling*, 25(10): 887–900.

Chau, K.W. and Cheng, C.T. (2002). "Real-time prediction of water stage with artificial neural network approach", *Lecture Notes in Artificial Intelligence*, 2557: 715–715.

Chau, K.W. and Jiang, Y.W. (2001). "3D numerical model for Pearl River Estuary", *Journal of Hydraulic Engineering, ASCE*, 127(1): 72–82.

Chau, K.W. and Jiang, Y.W. (2002). "Three-dimensional pollutant transport model for the Pearl River Estuary", *Water Research*, 36(8): 2029–2039.

Chau, K.W. and Jin, H.S. (1995). "Numerical solution of two-layer, two-dimensional tidal flow in a boundary-fitted orthogonal curvilinear co-ordinate system", *International Journal for Numerical Methods in Fluids*, 21(11): 1087–1107.

Chau, K.W. and Jin, H.S. (1998)."Eutrophication model for a coastal bay in Hong Kong", *Journal of Environmental Engineering, ASCE*, 124(7): 628–638.

Chau, K.W. and Jin, H.S. (2002). "Two-layered, 2D unsteady eutrophication model in boundary-fitted coordinate system", *Marine Pollution Bulletin*, 45(1–12): 300–310.

Chau, K.W. and Lee, J.H.W. (1991a). "Mathematical modelling of Shing Mun river network", *Advances in Water Resources*, 14(3): 106–112.

Chau, K.W. and Lee, J.H.W. (1991b). "A microcomputer model for flood prediction with applications", *Microcomputers in Civil Engineering*, 6(2): 109–121.

Chau, K.W. and Ng, V. (1996). "A knowledge-based expert system for design of thrust blocks for water pipelines in Hong Kong", *Journal of Water Supply Research and Technology – Aqua*, 45(2): 96–99.

Chau, K.W. and Yang, W.W. (1992). "A knowledge-based expert system for unsteady open channel flow", *Engineering Applications of Artificial Intelligence*, 5(5): 425–430.

Chau, K.W. and Yang, W.W. (1993). "Development of an integrated expert system for fluvial hydrodynamics", *Advances in Engineering Software*, 17(3): 165–172.

Chau, K.W. and Yang, W.W. (1994). "Structuring and evaluation of VP-Expert based knowledge bases", *Engineering Applications of Artificial Intelligence*, 7(4): 447–454.

Chau, K.W. and Yang, W.W. (1996). "Knowledge acquisition and representation for unsteady open channel flow", *Journal of Intelligent Systems*, 6(3–4): 221–237.

Chau, K.W. and Zhang, X.N. (1995). "An expert system for flow routing in a river network", *Advances in Engineering Software*, 22(3): 139–146.

Chau, K.W., Chuntian, C., and Li, C.W. (2002). "Knowledge management system on flow and water quality modeling", *Expert Systems with Applications*, 22(4): 321–330.

Chau, K.W., Jin, H.S., and Sin, Y.S. (1996). "A finite difference model of 2-D tidal flow in Tolo Harbour, Hong Kong", *Applied Mathematical Modelling*, 20(4): 321–328.

Chau, K.W., Lee, J.H.W., and Zienkiewicz, O.C. (1991). "Tidal transport calculations with the characteristic Galerkin method", in J.H.W. Lee and Y.K. Cheung, eds., *Environmental Hydraulics*, Vol. 2, A.A. Rotterdam, Balkema, pp. 1029–1034.

Chen, C.-F., Chen, Y.-C., and Lin, J.-Y. (2008). "Determination of optimal water resource management through a fuzzy multiobjective programming and genetic algorithm: Case study in Kinman, Taiwan", *Practice Periodical of Hazardous, Toxic, and Radioactive Waste Management*, 12(2): 86–95.

Chen, C.S., Liu, H.D., and Beardsley, R.C. (2003). "An unstructured grid, finite-volume, three-dimensional, primitive equations ocean model: Application to coastal ocean and estuaries", *Journal of Atmospheric and Oceanic Technology*, 20(1): 159–186.

Chen, L. (2003). "A study of applying genetic programming to reservoir trophic state evaluation using remote sensor data", *International Journal of Remote Sensing*, 24(11): 2265–2275.

Chen, Q. and Mynett, A.E. (2003). "Integration of data mining techniques and heuristic knowledge in fuzzy logic modelling of eutrophication in Taihu Lake", *Ecological Modelling*, 162(1–2): 55–67.

Chen, Q., Morales-Chaves, Y., Li, H., and Mynett, A.E. (2006). "Hydroinformatics techniques in eco-environmental modelling and management," *Journal of Hydroinformatics*, 8(4): 297–316.

Chen, S.H., Jakeman, A.J., and Norton, J.P. (2008). "Artificial intelligence techniques: An introduction to their use for modelling environmental systems," *Mathematics and Computers in Simulation*, 78(2–3): 379–400.

Chang, N.B., Chen, H.W., and Ning, S.K. (2001). "Identification of river water quality using the fuzzy synthetic evaluation approach", *Journal of Environmental Management*, 63(3): 293–305.

Cheng, C.T., Ou, C.P., and Chau, K.W. (2002). "Combining a fuzzy optimal model with a genetic algorithm to solve multiobjective rainfall-runoff model calibration", *Journal of Hydrology*, 268(1–4): 72–86.

Cheng, C.T., Chau, K.W., Sun, Y.G., and Lin, J.Y. (2005). "Long-term prediction of discharges in Manwan Reservoir using artificial neural network models", *Lecture Notes in Computer Science*, 3498: 1040–1045.

Cheng, R.T., Casulli, V., and Milford, S.N. (1984). "Eulerian-Lagrangian solution of the convection-dispersion equation in natural coordinates", *Water Resources Research*, 20(7): 944–952.

Cho, J.H., Sung, K.S., and Ha, S.R. (2004). "A river water quality management model for optimising regional wastewater treatment using a genetic algorithm", *Journal of Environmental Management*, 73(3): 229–242.

Choi, K.W., Lee, J.H.W., and Cheung, Y.K. (1989). "A numerical study of tidal flushing in a typhoon shelter", *Proceeding of the Fourth Asian Fluid Mechanics Congress*, Hong Kong, The University of Hong Kong, Volume 1, pp. B32–B35.

Choi, S.U., Kang, H., and Joung, Y. (2002). "Layer-averaged model for density currents spreading on a sloped surface", in *Computational Methods in Water Resources, Book Series: Developments in Water Science*, Amsterdam, Computational Mechanics Publications, 47: 1693–1700.

Choudhury, S., Mitra, S., and Chakraborty, H. (2004). "A connectionist approach to thunderstorm forecasting", *Annual Conference of the North American Fuzzy Information Processing Society – NAFIPS* 1: 330–334.

Chung, T.J. and Chiou, J.N. (1976). "Analysis of unsteady compressible boundary layer flow via finite elements", *Computers & Fluids*, 4: 1–12.

Clayton, B.D. (1985). *ART-Programming Tutorial, Vols 1–3*, Los Angeles, CA, Inference Corporation.

Clerc, M. and Kennedy, J. (2002). "The particle swarm – explosion, stability, and convergence in a multidimensional complex space", *IEEE Transactions on Evolutionary Computation*, 6(1): 58–73.

Cunge, J. (1989). "Review of recent developments in river modelling", in Falconer, R.A., Goodwin, P., and Matthew, R.G.S., eds., *Hydraulic and Environmental Modelling of Coastal, Estuarine and River Waters*, Aldershot, Avebury Technical, pp. 393–404.

Cunge, J.A., Holly, F.M., and Verwey, A. (1980). *Practical Aspects of Computational River Hydraulics*, Boston, MA, Pitman.

Dai, J.J., Lorenzato, S., and Rocke, D.M. (2004). "A knowledge-based model of watershed assessment for sediment", *Environmental Modelling and Software*, 19(4): 423–433.

Daoud, A.H., Rakha, K.A., and Abul-azm, A.G. (2008). "A two-dimensional finite volume hydrodynamic model for coastal areas: Model development and validation", *Ocean Engineering*, 35(1): 150–164.

Davies, A.M, Jones, J.E., and Xing, J. (1995). "Review of recent developments in tidal hydrodynamic model. II: Turbulence energy models", *Journal of Hydraulic Engineering, ASCE*, 123(4): 293–302.

Davis, R., *et al.* (1981). "The Dipmeter Advisor: interpretation of geologic signals", *Proceedings of the 7th International Joint Conference on Artificial Intelligence*, Vancouver, BC, pp. 846–849.

Deardorf, J.W. and Peterson, E.W. (1980). "The boundary-layer growth equation with Reynolds averaging", *Journal of the Atmospheric Sciences*, 37(6): 1405–1409.

Donea, J. (1984). "A Taylor-Galerkin method for convective transport problems", *International Journal of Numerical Methods in Engineering*, 20: 101–119.

Dronkers, J.J. (1964). *Tidal Computations in Rivers and Coastal Waters*, Amsterdam, North Holland Publishing Company.

Dronkers, J.J. (1969). "Tidal computations for rivers, coastal areas, and seas", *Proceedings of ASCE, Hydraulics Division*, HY1: 29–77.

Duan, W.Y., Xu, G.D., and Wu, G.X. (2009). "Similarity solution of oblique impact of wedge-shaped water column on wedged coastal structures", *Coastal Engineering*, 56(4): 400–407.

Duda, R.O. *et al.* (1979). *A Computer-Based Consultant for Mineral Exploration, Final Report SRI Project 6415*, Menlo Park, CA, SRI International.

Dublin Core Metadata Initiative (2009). *Dublin Core Metadata Element Set Version 1.1*, online, available at: http://dublincore.org/documents/dces/

Dym, C.L. (1987). "Implementation issues in the building of expert systems", in M.L. Maher, ed., *Expert Systems for Civil Engineers: Technology and Application*, New York, ASCE, pp. 9–18.

Dym, C.L. and Levitt, R.E. (1991). *Knowledge-Based Systems in Engineering*, New York, McGraw-Hill.

Elfeki, A.M.M., Uffink, G.J.M., and Lebreton, S. (2007). "Simulation of solute transport under oscillating groundwater flow in homogeneous aquifers", *Journal of Hydraulic Research*, 45(2): 254–260.

Engelmore, R. and Morgan, T. (1988). *Blackboard Systems*, Wokingham, Addison-Wesley.

EPD (1999). *Marine Water Quality in Hong Kong: Results for 1998 from the Marine Monitoring Program of the Environmental Protection Department*, Hong Kong Government, Hong Kong.

Erman, L.D., London, P.E., and Fickas, S.F. (1981). "The design and example use of Hearsay-III", *Proceedings of the 7th International Conference on AI*, San Francisco, CA, Vancouver, Morgan Kaufmaan Publishers Inc, pp. 409–415.

Escribano, R., Rosales, S.A., and Blanco, J.L. (2004). "Understanding upwelling circulation off Antofagasta (northern Chile): A three-dimensional numerical-modeling approach", *Continental Shelf Research*, 24(1): 37–53.

Falconer, R.A., Lin, B., Harris, E.L., and Wilson, C.A.M.E., eds. (2002). *Proceedings of the Fifth International Conference on Hydroinformatics*, 1–5 July 2002, Cardiff and Bristol, UK, IWA Publishing.

Fazio, P., Bedard, C., and Gowri, K. (1988). "Knowledge-based system development tools for processing design specifications", *Microcomputers in Civil Engineering*, 3(4): 333–344.

Fdez-Riverola, F. and Corchado, J.M. (2003). "CBR based system for forecasting red tides", *Knowledge-Based Systems*, 16 (5–6): 321–328.

Fenves, S.J. (1989). "Potentials of artificial intelligence and expert systems in computational mechanics", in A.K. Noor and J.T. Oden, eds., *State-of-the-Art Surveys on Computational Mechanics*, New York, ASME.

Fernandes, E.H.L., Dyer, K.R., Moller, O.O., *et al.* (2002) "The Patos Lagoon hydrodynamics during an El Niño event (1998)", *Continental Shelf Research*, 22(11–13): 1699–1713.

Finlayson, B.A. (1972). *The Method of Weighted Residuals and Variational Principles*, New York, Academic Press.

Fischer, H.B. and List, E.J. (1979). *Mixing in Inland and Coastal Waters*, New York, Academic Press.

Fitch, J.P., Lehman, S.K., Dowla, F.U., Lu, S.K., Johansson, E.M., and Goodman, D.M. (1991). "Ship wake detection procedure using conjugate gradient trained Artificial Neural Network", *IEEE Transactions on Geosciences and Remote Sensing*, 9(5): 718–725.

Forgy, C.L. (1981). *OPS5 User's Manual, Technical Report CMU-CS-81-135*, Pittsburgh, PA, Carnegie-Mellon University.

Friederich, S. and Gargano, M. (1989). *Expert Systems Design and Development Using VP-Expert*, New York, Wiley.

Galerkin, B.G. (1915). "Series occurring in some problems of elastic stability of rods and plates", *Engineering Bulletin*, 19: 897–908.

Garrett, J.H. Jr. (1994). "Where and why artificial neural networks are applicable in civil engineering", *Journal of Computing in Civil Engineering, ASCE*, 8(2): 129–130.

Gaschnig, J., Reboh, R., and Reiter, J. (1981). *Development of a Knowledge-based System for Water Resources Problems, SRI Project 1619*, Menlo Park, CA, SRI International.

Ghostine, R., Kesserwani, G., Vazquez, J., Riviere, N., Ghenaim, A., and Mose, R., (2009). "Simulation of supercritical flow in crossroads: Confrontation of a 2D and 3D numerical approaches to experimental results", *Computers & Fluids*, 38(2): 425–432.

Gilbert, R.W., Zedler, E.A., Grilli, S.T., and Street, R.L. (2007). "Progress on nonlinear-wave-forced sediment transport simulation", *IEEE Journal of Oceanic Engineering*, 32(1): 236–248.

Goldberg, D.E. (1989). *Genetic Algorithms for Search, Optimization and Machine Learning*, Reading, MA, Addison-Wesley.

Goldberg, D.E. and Deb, K. (1991). "A comparative analysis of selection schemes used in genetic algorithms", in Rawlins, G.J.E., ed., *Foundations of Genetic Algorithms*, San Mateo, CA, Morgan Kaufmann, pp. 69–93.

Goldberg, D.E. and Kuo, C.H. (1987). "Genetic algorithms in pipeline optimization", *Journal of Computing in Civil Engineering, ASCE*, 1(2): 128–141.

Goubesville, P., Cunge, J., Guinot, V., and Liong, S.Y., eds. (2006). *Proceedings of the Seventh International Conference in Hydroinformatics*, Nice, France, 4–7 September 2006, Chennai, India, Research Publishing Services.

Gournelos, T., Vassilopoulos, A., and Evelpidou, N. (2002). "Erosional processes in the north-eastern part of Attica (Oropos coastal zone) using web-GIS and soft computing technology", *Management Information Systems*, Ashurst, Wessex Institute of Technology, Ashurst Lodge, 415–424.

Guangdong Province Yearbook Editorial Committee (1996). *Guangdong Province Yearbook*, Guangzhou, Guangdong Province.

Guarino N. (1997). "Understanding, building, and using ontologies: A commentary to using explicit ontologies in KBS Development by van Heijst, Schreiber, and Wielinga", *International Journal of Human and Computer Studies*, 46(2–3): 293–310.

Habib, E. and Meselhe, E.A. (2006). "Stage-discharge relations for low-gradient tidal streams using data-driven models", *Journal of Hydraulic Engineering*, 132(5): 482–492.

Hagan, M.T., Demuth, H.B., and Beale, M. (1996). *Neural Network Design*, Boston, MA, London, PWS Publishing Company.

Hagen, S.C., Westerink, J.J., and Kolar, R.L. (2001). "One-dimensional finite element grids based on a localized truncation error analysis", *International Journal for Numerical Methods In Fluids*, 32(2): 241–261.

Haney, R.L. (1990) "Notes and correspondence: On the pressure gradient force over steep topography in sigma coordinate ocean model", *Journal of Physical Oceanography*, 21(4): 610–619.

Harleman, D.R.F. and Lee, C.H. (1969). *The Computation of Tides and Currents in Estuaries and Canals*, Technical Bulletin No. 16, Committee on Tidal Hydraulics, US Army Corps of Engineers.

Hayes-Roth, B. (1983). *The Blackboard Architecture: A General Framework for Problem Solving*, HPP 83-30, Department of Computer Science, Palo Alto, CA, Stanford University.

Haykin, S. (1999). *Neural Networks: A Comprehensive Foundation*, 2nd edition, Upper Saddle River, NJ, Prentice Hall.

Holland, J.H. (1975). *Adaptation in Natural and Artificial Systems*, Ann Arbor, MI, University of Michigan Press.

Holland, J.H. (1992). "Genetic algorithms", *Scientific American*, 267(1): 66–72.

Holly, F.M. Jr. and Preissmann, A. (1977) "Accurate calculation of transport in two dimensions", *Journal of Hydraulic Engineering, ASCE*, 103(11): 1259–1277.

Hong Kong Royal Observatory (1994). *Hong Kong Tide Table*, Hong Kong, Hong Kong Royal Observatory.

Huang, S. and Lu, Q.M. (1995). *Estuarine Dynamics*, Beijing, The Water Conservancy and Electricity Press, pp. 11–20.

Huang, W. and Foo, S. (2002). "Neural network modeling of salinity in Apalachicola River", *Water Resources Research*, 31(1): 2517–2530.

Hydrographic Department (1975). *Admiralty Tidal Stream Atlas Hong Kong*, Hong Kong, Hydrographic Department.

Information Builders, Inc. (1995). *LEVEL5 OBJECT® for Microsoft® Windows™ Reference Guide and Getting Start Guide*, Release 3.6, New York.

IntelliCorp (1986). *KEE Software Development User Manual*, 3rd ed.

Iowa Institute of Hydraulic Research (2000). *Proceedings of the Fourth International Conference on Hydroinformatics*, 23–27 July 2000, Cedar Rapids, IA.

Ip, S.F. and Wai, H.G. (1990). *An Application of Harmonic Method to Tidal Analysis and Prediction in Hong Kong*, Royal Observatory, Hong Kong Technical Note (local) No. 55.

Jang, J.S.R. (1993). "ANFIS: Adaptive-Network-based Fuzzy Inference Systems", *IEEE Transactions on Systems, Man, and Cybernetics*, 23(3): 665–685.

Jang, J.S.R. and Sun C.T. (1995). "Neuro-fuzzy modeling and control", *Proceedings of the IEEE*, 83(3): 378–406.

Jeong, K.S., Joo, G.J., Kim, H.W., Ha, K., and Recknagel, F. (2001). "Prediction and elucidation of algal dynamics in the Nakdong River (Korea) by means of a recurrent artificial neural network", *Ecological Modelling*, 146(1–3): 115–129.

Jeong, K.S., Kim, D.K., Whigham, P., and Joo, G.J. (2003). "Modelling *Microcystis aeruginosa* bloom dynamics in the Nakdong River by means of evolutionary computation and statistical approach", *Ecological Modelling*, 161(1–2): 67–78.

Jones, J.E. and Davies, A.M. (2007). "A high-resolution finite element model of the M2, M4 M6, S2, N2, K1 and O1 tides off the west coast of Britain", *Ocean Modelling*, 19(1–2): 70–100.

Kalra, R. and Deo, M.C. (2007). "Genetic programming for retrieving missing information in wave records along the west coast of India", *Applied Ocean Research*, 29(3): 99–111.

Kangari, R., and Boyer, L.T. (1987). "Knowledge-based systems and fuzzy sets in risk management", *Microcomputers in Civil Engineering*, 2(4): 272–283.

Karamperidou, C., Karamperidou, E., and Katsifarakis, K.L. (2007). "Seawater intrusion into the aquifer of Eleftherae-N. Peramos, Kavala, Greece", *WIT Transactions on Ecology and the Environment*, 104: 3–10.

Karul, C., Soyupak, S., Cilesiz, A.F., Akbay, N., and Germen, E. (2000). "Case studies on the use of neural networks in eutrophication modelling", *Ecological Modelling*, 134(2–3): 145–152.

Keen, T.R., Bentley, S.J., Vaughan, W.C., *et al.* (2004). "The generation and preservation of multiple hurricane beds in the northern Gulf of Mexico", *Marine Geology*, 210(1–4): 79–105.

Kennedy, J. (1997). "The particle swarm: Social adaptation of knowledge", *Proceedings of the 1997 International Conference on Evolutionary Computation*. Indianapolis, pp. 303–308.

Kennedy, J. and Eberhart, R. (1995). "Particle swarm optimization", *Proceedings of the 1995 IEEE International Conference on Neural Networks*, Perth, pp. 1942–1948.

Kliem, N., Nielsen, J.W., and Huess, V. (2006). "Evaluation of a shallow water unstructured mesh model for the North Sea-Baltic Sea", *Ocean Modelling*, 15(1–2): 124–136.

Knight, B. and Petridis, M. (1992). "Flowes: An intelligent computational fluid dynamics system", *Engineering Applications of Artificial Intelligence*, 5(1): 51–58.

Kodama, T., Wang, S.S.Y., and Kawahara, M. (2001). "Model verification on 3D tidal current analysis in Tokyo Bay", *International Journal for Numerical Methods in Fluids*, 22(1): 43–66.

Kot, S.C. and Hu, S.L. (1995). "Water flows and sediment transport in Pearl River estuary and wave in South China sea near Hong Kong", in *Coastal Infrastructure Development in Hong Kong – a Review*, Hong Kong Government, pp. 13–32.

Koulouriotis D.E., Diakoulakis I.E., Emiris D.M., and Zopounidis C.D. (2005). "Development of dynamic cognitive networks as complex systems approximators: Validation in financial time series", *Applied Soft Computing*, 5(2): 157–179.

Koza, J. (1992). *Genetic Programming: On the Programming of Computers by Natural Selection*, Cambridge, MA, MIT Press.

Kralisch, S., Fink, M., Flügel, W.A., and Beckstein, C. (2003). "A neural network approach for the optimisation of watershed management", *Environmental Modelling & Software*, 18(8–9): 815–823.

Kuipers, J. and Vreugdenhill, C.B. (1973). *Calculations of Two-Dimensional Horizontal Flow*, Delft Hydraulics Laboratory, Research Report S-163-I.

Kumar, B. (1995). *Knowledge Processing for Structural Design*, Topics in Engineering, Volume 25, Southampton, Computational Mechanics.

Kumar, B.P., Kumar, R.R., Dube, S.K., Rao, A.D., Murty, T., Gangopadhyay, A., and Chaudhuri, A. (2008). "Tsunami early warning system: An Indian Ocean perspective", *Journal of Earthquake and Tsunami*, 2(3): 197–226.

Lachaume, C., Biausser, B., Grilli, S.T., Fraunié, P., and Guignard, S. (2003). "Modeling of breaking and post-breaking waves on slopes by coupling of BEM and VOF Methods", *Proceedings of the International Offshore and Polar Engineering Conference*, Golden, CO, International society of Offshore and Polar Engineers, pp. 1698–1704.

Laguzzi, M.M., Stelling, G.S., and de Brujin, K. (2001). "A fully coupled 1D & 2D system specially suited for flooding simulation", *Proceedings of 29th Annual Congress of the International Association of Hydraulic Engineering and Research (IAHR)*, Beijing, People's Republic of China, pp. 36–40.

Lee, J.H.W. and Arega, F. (1999). "Eutrophication dynamics of Tolo Harbour, Hong Kong", *Marine Pollution Bulletin*, 39(1–12): 187–192.

Lee, J.H.W., Peraire, J., and Zienkiewicz, O.C. (1987). "The characteristic-Galerkin method for advection-dominated problems: An assessment", *Computer Methods in Applied Mechanics and Engineering*, 61: 359–369.

Lee, J.H.W., Huang, Y., Dickman, M., and Jayawardena, A.W. (2003). "Neural network modelling of coastal algal blooms", *Ecological Modelling*, 159(2–3): 179–201.

Leendertse, J.J. (1967). *Aspects of a Computational Model for Long-Period Water-Wave Propagation*, RAND Corporation, Memo RM-5294-PR.

Leendertse, J.J. and Crittion, E.C. (1971). *A Water Quality Simulation Model for Well-Mixed Estuaries and Coastal Seas Computation Procedures*, R-708-NYC.11, New York, Rand Corporation.

Leendertse, J.J., Alexander, R.C., and Liu, S.K. (1973). *A Three-Dimensional Model for Estuaries and Coastal Seas, Volume I. Principles of Computation*, R-1417-OWRR, Santa Monica, CA, Rand Corporation.

Lek, S. and Guegan, J.F. (1999). "Artificial neural networks as a tool in ecological modelling: An introduction", *Ecological Modelling*, 120(2–3): 65–73.

Level Five Research (1986). *INSIGHT 2+ Reference Manual, Version 1.0*, Indialantic, FL, Level Five Research, Inc.

Li, C.W. and Lee, J.H.W. (1985). "Continuous source in tidal flow: A numerical study of the transport equation", *Applied Mathematical Modelling*, 9: 281–288.

Lin, B.L. and Chandler-Wilde, S.N. (1996). "A depth-integrated 2D coastal and estuarine model with conformal boundary-fitted mesh generation", *International Journal for Numerical Methods in Fluids*, 23(8): 819–846.

Lin, B., Syed, M., and Falconer, R.A. (2008). "Predicting faecal indicator levels in estuarine receiving waters: An integrated hydrodynamic and ANN modelling approach", *Environmental Modelling and Software*, 23(6): 729–740.

Lindsay, R.K., Buchanan, B.G., Feigenbaum, E.A., and Lederberg, J. (1980). *Applications of Artificial Intelligence for Organic Chemistry – The DENDRAL Project*, New York, McGraw-Hill.

Liong, S.Y., Phoon, K.K., and Babovic, V., eds. (2004) *Proceedings of the Sixth International Conference on Hydroinformatics*, 21–24 June 2004, Singapore.

Liou, S.M., Lo, S.L., and Hu, C.Y. (2003). "Application of two-stage fuzzy set theory to river quality evaluation in Taiwan", *Water Research*, 37(6): 1406–1416.

Liu, C., Liu, X.P., and Jiang, C.B. (2005). "Numerical simulation of wave field near submerged bars by PLIC-VOF model", *China Ocean Engineering*, 19 (3), 509–518.

Lo, S.H. (1991). "Automatic mesh generation and adaptation by using contours", *International Journal for Numerical Methods in Engineering*, 31(4): 689–707.

Lo, S.H. (1992). "Generation of high-quality gradation finite element mesh", *Engineering Fracture Mechanics*, 41(2): 191–202.

Lo, S.H. and Lee, C.K. (1994). "Generation of gradation meshes by the background grid technique", *Computers & Structures*, 50(1): 21–32.

Lo, S.P. (2003). "An adaptive-network based fuzzy inference system for prediction of workpiece surface roughness in end milling", *Journal of Materials Processing Technology*, 142(3): 665–675.

Loia, V., Sessa, S., Staiano, A., and Tagliaferri, R. (2000). "Merging fuzzy logic, neural networks, and genetic computation in the design of a decision-support system", *International Journal of Intelligent Systems*, 15(7): 575–594.

Londhe, S.N. (2008). "Development of wave buoy network using soft computing techniques", *OCEANS'08 MTS/IEEE Kobe-Techno-Ocean'08 – Voyage toward the Future, OTO'08*, art. no. 4530913.

Lu, Q.M. (1997). *Three-Dimensional Modeling of Hydrodynamics and Sediment Transport with Parallel Algorithm*, thesis for Ph.D., Hong Kong Polytechnic University.

Lynch, D.R. and Gray, W.G. (1978) "Analytical solutions for computer flow model testing", *Journal of Hydraulic Engineering, ASCE*, 104, HY10, 1409–1428.

Maher, M.L., ed. (1987). *Expert Systems for Civil Engineers: Technology and Application*, New York, American Society of Civil Engineers.

Maher, M.L., Fenves, S.J., and Garrett, J.H. (1988). "Expert systems for structural design", Adeli, H. ed., *Expert Systems in Construction and Structural Engineering*, New York, Chapman & Hall, pp. 85–121.

Mahjoobi, J. and Adeli Mosabbeb, E. (2009). "Prediction of significant wave height using regressive support vector machines", *Ocean Engineering*, 36(5): 339–347.

Mahjoobi, J., Etemad-Shahidi, A., and Kazeminezhad, M.H. (2008). "Hindcasting of wave parameters using different soft computing methods", *Applied Ocean Research*, 30(1): 28–36.

Maier, H.R. and Dandy, G.C. (1997). "Modelling cyanobacteria (blue-green algae) in the River Murray using artificial neural networks", *Mathematics and Computers in Simulation*, 43(3–6): 377–386.

Maier, H.R. and Dandy, G.C. (2000). "Neural networks for the predication and forecasting of water resources variables: A review of modelling issues and applications", *Environmental Modelling & Software*, 15(1): 101–124.

Maier, H.R., Dandy, G.C., and Burch, M.D. (1998). "Use of artificial neural networks for modelling cyanobacteria *Anabaena* spp. in the River Murray, South Australia", *Ecological Modelling*, 105(2–3): 257–272.

Maier, H.R., Morgan, N., and Chow, C.W.K. (2004). "Use of artificial neural networks for predicting optimal alum doses and treated water quality parameters", *Environmental Modelling & Software*, 19(5): 485–494.

Maier, H.R., Sayed, T., and Lence, B.J. (2001). "Forecasting cyanobacterium *Anabaena* spp. in the River Murray, South Australia, using B-spline neurofuzzy models", *Ecological Modelling*, 146(1–3): 85–96.

Mamdani, E.H. and Assilian S. (1975). "An experiment in linguistic synthesis with a fuzzy logic controller", *International Journal of Man-Machine Studies*, 7(1): 1–13.

Marsili-Libelli, S. (2004). "Fuzzy prediction of the algal blooms in the Orbetello lagoon", *Environmental Modelling & Software*, 19(9): 799–808.

Marinov, D., Norro, A., and Zaldivar, J.M. (2006). "Application of COHERENS model for hydrodynamic investigation of Sacca di Goro coastal lagoon (Italian Adriatic Sea shore)", *Ecological Modelling*, 193(1–2): 52–68.

Martin, J.L., McCutcheon, S.C., and Schottman, R.W. (1999). *Hydrodynamics and Transport for Water Quality Modeling*, Boca Raton, FL, Lewis Publishers.

Masters, T. (1993). *Practical Neural Networks Recipes C++*, San Diego, Academic Press.

Mattocks, C. and Forbes, C. (2008). "A real-time, event-triggered storm surge forecasting system for the state of North Carolina", *Ocean Modelling*, 25(3–4): 95–119.

McDermott, J. (1980). *R1: A Rule-Based Configurer of Computer Systems, Technical Report CMU-CS-80-119*, Department of Computer Science, Pittsburgh, PA, Carnegie-Mellon University.

Mellor, G.L. (1996). *User's Guide for a Three-Dimensional, Primitive Equation, Numerical Ocean Model*, New Jersey, Princeton University Rep.

Mellor, G.L., Oey, L.Y., and Ezer, T. (1997). "Sigma coordinate pressure gradient errors and the seamount problem", *Journal of Atmospheric and Oceanic Technology*, 15(5): 1122–1131.

Mestres, M., Sierra, J.P., Sanchez-Arcilla, A., *et al.* (2003). "Modelling of the Ebro River plume: Validation with field observations", *Scientia Marina*, 67(4): 379–391.

Mitchell, T.M. (1997). *Machine Learning*, New York, McGraw-Hill.

Mohan, S. (1990). "Status of expert systems education in civil engineering", *International Journal of Applied Engineering Education*, 6(2): 131–143.

Moller, M.F. (1993). "A scaled conjugate gradient algorithm for fast supervised learning", *Neural Networks*, 6(4): 523–533.

Morton, B. (1988). "Editorial: Hong Kong's first marine disaster", *Marine Pollution Bulletin*, 19(7): 299–300.

Moller, M.F. (1993). "A scaled conjugate gradient algorithm for fast supervised learning", *Neural Networks*, 6(4): 523–533.

Moore, T., Morris, K., Blackwell, G., Gibson, S., and Stebbing, A. (1999). "An expert system for integrated coastal zone management: A geomorphological case study", *Marine Pollution Bulletin*, 37(3–7): 361–370.

Motik, B., Maedche, A., and Volz, R. (2002). "A conceptual modeling approach for semantics-driven enterprise applications", *Lecture Notes in Artificial Intelligence*, 2519: 1082–1099.

Müller, A., ed. (1996). *Proceedings of the Second International Conference on Hydroinformatics*, 9–13 September 1996, Zurich, Switzerland, A.A. Balkema Publishers.

Muttin, F. (2008). "Oil spill boom modelling by the finite-element method", *Water Pollution IX, Book Series: WIT Transactions on Ecology and the Environment*, 111: 383–392.

Muttil, N. and Chau, K.W. (2007). "Machine-learning paradigms for selecting ecologically significant input variables", *Engineering Applications of Artificial Intelligence*, 20(6): 735–744.

Muttil, N., Lee, J.H.W., and Jayawardena, A.W. (2004). "Real-time prediction of coastal algal blooms using genetic programming", *Proceedings of the Sixth International Conference on Hydroinformatics*, S.Y. Liong, K.K. Phoon and V. Babovic, eds., 21–24 June 2004, Singapore, pp. 890–897.

Nayak, P.C., Sudheer, K.P., Rangan, D.M., and Ramasastri, K.S. (2004). "A neuro-fuzzy computing technique for modeling hydrological times series", *Journal of Hydrology*, 291(1–2): 52–66.

Nemati, H.R., Steiger, D.M., Iyer, L.S., and Herschel, R.T. (2002). "Knowledge warehouse: An architectural integration of knowledge management, decision support, artificial intelligence and data warehousing", *Decision Support Systems*, 33(2): 143–161.

Nii, H.P. and Aiello, N. (1979). "AGE (Attempt to Generalize): A knowledge-based program for building knowledge-based programs", *Proceedings of the Sixth International Joint Conference on AI*, Tokyo, pp. 645–655.

Norton, W.R. and King, I.P. (1976). *User's Guide and Operating Instructions for the Computer Program RMA-2*, Lafayette, CA, Resource Management Associates.

Oden, J.T. (1972). *Finite Elements of Nonlinear Continua*, New York, McGraw-Hill.

Odgaard, A.J. (2001). "Trends and current developments in hydraulic engineering", *Workshop on Integrated Water Resources Management*, Hsinchu, Taiwan, National Chiao Tung University, 15–16 October 2001, pp. 1–7.

Oey, L.Y., Mellor, G.L., and Hires, R.I. (1985). "A three-dimensional simulation of the Hudson Raritan estuary. Part I: Description of model and model simulation", *Journal of Physical Oceanography*, 15(12): 1693–1709.

Okubo, A. and Karweit, M. (1969). "Diffusion from a continuous source in a uniform shear flow", *Limnology and Oceanography*, 14(4): 514–520.

Pang, Y. and Li, X.L. (1998). "Study of pollutants passing through the four east outlets of Pearl river delta to Lingding sea", *Proceedings, Workshop on Hydraulics of the Pearl River Estuary*, Y.S, Li, ed., Hong Kong, Hong Kong Polytechnique University, pp. 85–98.

Pape, L., Ruessink, B.G., Wiering, M.A., and Turner, I.L. (2007). "Recurrent neural network modeling of nearshore sandbar behavior", *Neural Networks*, 20(4): 509–518.

Patel, D., Guganesharajah, K., and Thake, B. (2004). "Modelling diatom growth in turbulent waters", *Water Research*, 38(11): 2713–2725.

Peraire, J., Zienkiewicz, O.C., and Morgan, K. (1986). "Shallow water problems: a general explicit formulation", *International Journal for Numerical Methods in Engineering*, 22: 547–574.

Pereira, G.C., and Ebecken, N.F.F. (2009). "Knowledge discovering for coastal waters classification", *Expert Systems with Applications*, 36(4): 8604–8609.

Perera, E.D.P., Jinno, K., Tsutsumi, A., and Hiroshiro, Y. (2008). "Development and verification of a three dimensional density dependent solute transport model for seawater intrusion", *Memoirs of the Faculty of Engineering, Kyushu University*, 68(2): 93–106.

Pinthong, P., Gupta, A.D., Babel, M.S., and Weesakul, S. (2009). "Improved reservoir operation using hybrid genetic algorithm and neurofuzzy computing," *Water Resources Management*, 23(4): 697–720.

Pinho, J.L.S., Vieira, J.M.P., and do Carmo, J.S.A. (2004). "Hydroinformatic environment for coastal waters hydrodynamics and water quality modelling", *Advances in Engineering Software*, 35(3–4): 205–222.

Ponnambalam, K., Karray, F., and Mousavi, S.J. (2002). "Minimizing variance of reservoir systems operations benefits using soft computing tools", *Fuzzy Sets and Systems*, 139(2): 451–461.

Pople, H.E. Jr. (1982). "Heuristic methods for imposing structure on ill-structured problems: The structure of medical diagnosis", in P. Szolovits, ed., *Artificial Intelligence in Medicine*, Boulder, CO, Westview Press.

Preis, A., and Ostfeld, A. (2008). "A coupled model tree-genetic algorithm scheme for flow and water quality predictions in watersheds," *Journal of Hydrology*, 349(3–4): 364–375.

Qi, J., Chen, C., Beardsley, R.C., Perrie, W., Cowles, G.W., and Lai, Z. (2009). "An unstructured-grid finite-volume surface wave model (FVCOM-SWAVE): Implementation, validations and applications", *Ocean Modelling*, 28(1–3): 153–166.

Quamrul, A.K.M. and Blumberg, A.F. (1999). "Three-dimensional model of Onondaga Lake, New York", *Journal of Hydraulic Engineering, ASCE*, 125(9): 912–923.

Ragas, A.M.J., Haans, J.L.M., and Leuven, R.S.E.W. (1997). "Selecting water quality models for discharge permitting", *European Water Pollution Control*, 7(5): 59–67.

Ranga Rao, A.V. and Sundaravadivelu, R. (1999). "A knowledge based expert system for design of berthing structures", *Ocean Engineering*, 26(7): 653–673.

Rayleigh, J.W.S. (1877). *Theory of Sound*, 1st ed., revised, Dover, New York, 1945.

Recknagel, F. (2001). "Applications of machine learning to ecological modelling", *Ecological Modelling*, 146(1–3): 303–310.

Recknagel, F., Bobbin, J., Whigham, P., and Wilson, H. (2002). "Comparative application of artificial neural networks and genetic algorithms for multivariate time-series modelling of algal blooms in freshwater lakes", *Journal of Hydroinformatics*, 4(2): 125–134.

Recknagel, F., French, M., Harkonen, P., and Yabunaka, K. (1997). "Artificial neural network approach for modelling and prediction of algal blooms", *Ecological Modelling*, 96(1–3): 11–28.

Recknagel, F., Fukushima, T., Hanazato, T., Takamura, N., and Wilson, H. (1998). "Modelling and prediction of phyto- and zooplankton dynamics in Lake Kasumigaura by artificial neural networks", *Lakes & Reservoirs: Research and Management*, 3(2): 123–133.

Recknagel, F., Petzoldt, T., Jaeke, O., and Krusche, F. (1994). "Hybrid expert system DELAQUA: A toolkit for water quality control of lakes and reservoirs", *Ecological Modelling*, 71(1–3): 17–36.

Richtmyer, R.D. and Morton, K.W. (1967). *Difference Methods for Initial-Value Problems*, 2nd edition, New York, Wiley Interscience.

Ritz, W. (1909). "Uber eine neue Methode zur Losung gewisser Variations-probleme der mathematischen Physik", *Journal für die Reine und Angewandte Mathematik*, 135(1): 1.

Roache, P. (1976). *Computational Fluid Dynamics*, Hermosa, Albuquerque, NM.

Rouhani, S. and Kangari, R. (1987). "Landfill site selection: A microcomputer expert system", *Microcomputers in Civil Engineering*, 2(1): 47–53.

Różyński, G. and Jansen, H. (2002). "Modeling nearshore bed topography with principal oscillation patterns", *Journal of Waterway, Port, Coastal and Ocean Engineering, ASCE*, 128(5): 202–215.

Rule Machines Corporation (1998). *Visual Rule Studio Developer's Guide*, Indialantic, FL, Rule Machines Corporation.

Rumelhart, D.E., Hinton, D.E., and Williams, R.J. (1986). "Learning internal representations by error propagation", in Rumelhart, D. and McClelland, J. eds., *Parallel Distributed Processing: Exploration in the Microstructure of Cognition*, Vol. 1, Cambridge, CA, MIT Press, pp. 318–362.

Rumelhart, D.E., Widrow, B., and Lehr, M.A. (1994). "The basic ideas in neural networks", *Communications of the ACM*, 37(3): 87–92.

Russell, S.O. and Campbell, P.F. (1996). "Reservoir operating rules with fuzzy programming", *Journal of Water Resources Planning and Management, ASCE*, 122(3): 165–170.

Rychener, M. (1988). "Research in expert systems for engineering design", in Rychener, M., ed., *Expert Systems for Engineering Design*, Boston, MA, Academic Press, pp. 1–33.

Sahooa, G.B., Raya, C., and Wadeb, H.F. (2005). "Pesticide prediction in ground water in North Carolina domestic wells using artificial neural networks", *Ecological Modelling*, 183: 29–46.

Sakr, K.M. and Hosain, M.U. (1989). "Applications of expert system tools in structural design", *Comp. in Civil Eng., Proc. 6th Conf., ASCE*.

Scardi, M. (2001). "Advances in neural network modelling of phytoplankton primary production", *Ecological Modelling*, 146(1–3): 33–45.

Schories, D., Pehlke, C., and Selig, U. (2009). "Depth distributions of *Fucus vesiculosus* L. and *Zostera marina* L. as classification parameters for implementing the European Water Framework Directive on the German Baltic coast", *Ecological Indicators*, 9(4): 670–680.

Serodes, J.B. and Rodriguez, M.J. (1996). "Predicting residual chlorine evolution in storage tanks within distribution systems: Application of a neural-network approach", *Journal of Water Supply Research and Technology – Aqua*, 45(2): 57–66.

Shahin, M.A., Maier, H.R., and Jaksa, M.B. (2002). "Predicting settlement of shallow foundations using neural networks", *Journal of Geotechnical and Geoenviromental Engineering, ASCE,* 128(9): 785–793.

Shortliffe, E.H. (1976). *Computer-Based Medical Consultations: MYCIN,* New York, American Elsevier.

Shrestha, B.S., Duckstein, L., and Stakhiv, E.Z. (1996). "Fuzzy rule-based modeling of reservoir operation", *Journal of Water Resources Planning and Management, ASCE,* 122(4): 262–269.

Shwe, T.T. and Adeli, H. (1991). "An expert system for earthquake design code processing", *Structural Engineering Review,* 3: 41–48.

Silvert, W. (1997). "Ecological impact classification with fuzzy sets", *Ecological Modelling,* 96(1–3): 1–10.

Simpson, P.K. (1990). *Artificial Neural Systems: Foundations, Paradigms, Applications and Implementations,* New York, Pergamon.

Sivakumar, B., Jayawardena, A.W., and Fernando, T.M.K.G. (2002). "River flow forecasting: Use of phase-space reconstruction and artificial neural networks approaches", *Journal of Hydrology,* 265(1–4): 225–245.

Smith, M. (1993). *Neural Networks for Statistical Modeling,* New York, Van Nostrand Reinhold.

Sriram, D., Maher, M.L., and Fenves, S.J. (1985). "Knowledge-based expert systems in structural design", *Computers & Structures,* 20(1–3): 1–9.

Stefik, M.J. (1979). "An examination of a frame-structured representation system", *Proceedings Sixth International Joint Conference on AI,* Tokyo, pp. 845–852.

Stefik, M.J. (1981). "Planning with constraints (MOLGEN 1)", *Artificial Intelligence,* 16(2): 111–139.

Sugeno, M. and Kang, G.T. (1988). "Structure identification of fuzzy model", *Fuzzy Sets and Systems,* 28(1): 15–33.

Sun Microsystems Inc. (2009). *JavaTM 2 Platform Enterprise Edition (J2EETM),* online, available at: http://java.sun.com/j2ee/

Takagi, T. and Sugeno, M. (1985). "Fuzzy identification of systems and its applications to modeling and control", *IEEE Transactions on Systems, Man, and Cybernetics,* 15(1): 116–132.

Tang, H.S., Keen, T.R., and Khanbilvardi, R. (2009). "A model-coupling framework for nearshore waves, currents, sediment transport, and seabed morphology", *Communications in Nonlinear Science and Numerical Simulation,* 14(7): 2935–2947.

Thacker, W.C. (1980). "A brief review of techniques for generating irregular computational grids", *International Journal for Numerical Methods in Engineering,* 15(9): 1335–1341.

Thirumalaiah, K. and Deo, M.C. (1998). "River stage forecasting using artificial neural networks", *Journal of Hydrologic Engineering, ASCE,* 3(1): 26–32.

Tokar, A.S. and Johnson, P.A. (1999). "Rainfall-runoff modeling using artificial neural networks", *Journal of Hydrologic Engineering, ASCE,* 4(3): 232–239.

Tsukamoto, Y. (1979). "An approach to fuzzy reasoning method", *Advances in Fuzzy Set Theory and Applications,* 137–149.

Tucciarelli, T. and Termini, D. (2000). "Finite-element modeling of floodplain flow", *Journal of Hydraulic Engineering, ASCE,* 126(6): 418–424.

Turner, M., Clough, R., Martin, H., and Topp, L. (1956). "Stiffness and deflection analysis of complex structures", *Journal of the Aerospace Sciences,* 23(9): 805–823.

Uddameri, V. and Honnungar, V. (2007). "Combining rough sets and GIS techniques to assess aquifer vulnerability characteristics in the semi-arid South Texas", *Environmental Geology*, 51(6): 931–939.

Umeyama, M. and Shintani, T. (2004). "Visualization analysis of runup and mixing of internal waves on an upper slope", *Journal of Waterway, Port, Coastal and Ocean Engineering*, 130(2): 89–97.

Universidad de Concepción (2009). *Proceedings of the Eighth International Conference on Hydroinformatics*, 12–16 January 2009, Concepción, Chile.

University of Karlsruhe. (2009). *The Karlsruhe Ontology (KAON) tool suite*, online, available at: http://kaon.semanticweb.org/

van Melle, W. (1979). "A domain independent production-rule system for consultation programs", *Proceedings of the 6th International Joint Conference on AI*, Tokyo, pp. 923–925.

Verwey, A., Minns, A.W., Babovic, V., and Maksimovic, C., eds., (1994). *Proceedings of the First International Conference on Hydroinformatics*, Rotterdam, The Netherlands, Balkema.

Wai, O.W.H., Chen, Y., and Li, Y.S. (2004). "A 3-D wave-current driven coastal sediment transport model", *Coastal Engineering Journal*, 46(4): 385–424.

Walters, R.A., Hanert, E., Pietrzak, J., and Le Roux, D.Y. (2009). "Comparison of unstructured, staggered grid methods for the shallow water equations", *Ocean Modelling*, 28(1–3): 106–117.

Wang, B.D. (2000). *Fuzzy Mathematical Methods for Long-Term Hydrological Prediction*, Dalian, Dalian University of Technology Press.

Wang, J.D. and Connor, J.J. (1975). *Mathematical Modelling of Near Coastal Circulation*, Report No. 200, Ralph M. Parsons Laboratory, MIT.

Wei, B., Sugiura, N., and Maekawa, T. (2001). "Use of artificial neural network in the prediction of algal blooms", *Water Research*, 35(8): 2022–2028.

Weiss, S.M. and Kulikowski, C.A. (1979). "EXPERT: a system for developing consultation models", *Proceedings of the 6th International Joint Conference on AI*, Tokyo, pp. 942–947.

Wen, W.Y., Zhang, G.X., and Du, W.C. (1994). "A study on water pollution in the Zhujiang estuary", *Environmental Research of Pearl River Delta Region*, Guangzhou, Guangdong Province Government, pp. 99–151.

Westwood, I.J. and Holz, K.P. (1986). "Automatic optimization of irregular triangular meshes for natural flow computations", *Proceedings of Second International Conference on Hydraulic Engineering Software*, Southampton, Computational Mechanics Publications, pp. 423–435.

Whitehead, P.G., Howard, A., and Arulmani, C. (1997). "Modelling algal growth and transport in rivers: A comparison of time series analysis, dynamic mass balance and neural network techniques", *Hydrobiologia*, 349: 39–46.

Wright, J.M. and Fox, M.S. (1983). *SRL/1.5 User Manual*.

Wu, J.H. (1986). "An unconditionally L_∞-stable method of fractional steps for numerical solution of convective diffusion problems", *Proceedings of the 10th International Conference on Numerical Methods in Fluid Dynamics*, Beijing, pp. 660–665.

Wu, J. and Chen, K. (1985). "A hybrid method of fractional steps with L_∞-stability for numerical modelling of harbours and bays", *Proceedings of the International Conference on Numerical and Hydraulic Modelling of Ports and Harbours*, Birmingham, England, pp. 101–109.

Wu, C.L., Chau, K.W., and Li, Y.S. (2008). "River flow prediction based on a distributed support vector regression", *Journal of Hydrology*, 358(1–2): 96–111.

Wu, C.L., Chau, K.W., and Li, Y.S. (2009). "Methods to improve neural network performance in daily flows prediction", *Journal of Hydrology*, 372(1–4): 80–93.

Wu, X.G., Shen, Y.M., and Zheng, Y.H. (2004). "Vertical 2D modeling of free surface flow with hydrodynamic pressure using SIMPLE arithmetic in sigma coordinates", *China Ocean Engineering*, 18(1): 79–92.

Wylie, C.R. (1975) *Advanced Engineering Mathematics*, 4th edition, Tokyo, McGraw-Hill.

Xu, F. L., Lam, K.C., Zhao, Z.Y., Zhan, W., Chen, Y.D., and Tao, S. (2004). "Marine coastal ecosystem health assessment: A case study of the Tolo Harbour, Hong Kong, China", *Ecological Modelling*, 173(4): 355–370.

Yabunaka, K., Hosomi, M., and Murakami, A. (1997). "Novel application of a back-propagation artificial neural network model formulated to predict algal bloom", *Water Science & Technology*, 36(5): 89–97.

Young, D.L., Lin, Q.H., and Murugesan, K. (2005). "Two-dimensional simulation of a thermally stratified reservoir with high sediment-laden inflow", *Journal of Hydraulic Research*, 43(4): 351–365.

Yu, L. and Righetto, A.M. (2001). "Depth-averaged turbulence k-e model and applications", *Advances in Engineering Software*, 32(5): 375–394.

Zadeh, L. (1994). "Soft computing and fuzzy logic", *IEEE Software*, 11: 48–56.

Zadeh, L.A. (1965). "Fuzzy sets", *Information and Control*, 8(3): 338–353.

Zadeh, L. and Kacprzyk, J. (1992). *Fuzzy Logic for the Management of Uncertainty*, New York, John Wiley.

Zou, R., Lung, W.S., and Guo, H. (2002). "Neural network embedded Monte Carlo approach for water quality modelling under input information uncertainty", *Journal of Computing in Civil Engineering, ASCE*, 16(2): 135–142.

Zou, R., Lung, W.-S., and Wu, J. (2007). "An adaptive neural network embedded genetic algorithm approach for inverse water quality modeling", *Water Resources Research*, 43(8), art. no. W08427.

Index